SpringerBriefs on Case Studies of Sustainable Development

Series Editors

Asit K. Biswas, Atizapán, Mexico
Cecilia Tortajada, Atizapán, Mexico

For further volumes:
http://www.springer.com/series/11889

Asit K. Biswas · Cecilia Tortajada
Andrea Biswas-Tortajada
Yugal K. Joshi *with* Aishvarya Gupta

Creating Shared Value

Impacts of Nestlé in Moga, India

With a Foreword by

Michael Porter and Mark Kramer

and an Epilogue by

Peter Brabeck-Letmathe and Paul Bulcke

Springer

Asit K. Biswas
Lee Kuan Yew School of Public Policy
National University of Singapore
Singapore

Cecilia Tortajada
Andrea Biswas-Tortajada
Third World Centre for Water Management
Atizapán, Estado de México
Mexico

Yugal K. Joshi
Delhi Division
Northern Railway
New Delhi
India

Aishvarya Gupta
Nirma University
Ahmedabad
India

ISSN 2196-7830 ISSN 2196-7849 (electronic)
ISBN 978-3-319-01462-3 ISBN 978-3-319-01463-0 (eBook)
DOI 10.1007/978-3-319-01463-0
Springer Cham Heidelberg New York Dordrecht London

Library of Congress Control Number: 2013946309

Printed on acid-free paper

Springer is part of Springer Science+Business Media (www.springer.com)

Foreword

Over the past 50 years, Nestlé has demonstrated the power of *Creating Shared Value* in Moga, India. The company's strategy to enhance competitive advantage and value for its shareholders has simultaneously created a prosperous local dairy cluster that has improved local health, education, and standards of living for tens of thousands of formerly impoverished and malnourished people. Any government, charity, or development agency would be proud of the social impact that Nestlé has created through the pursuit of shared value in Moga.

The concept of shared value focuses on the interdependence of business and society.[1] Companies create shared value when they employ policies and practices that create economic benefit for the business while creating social and environmental benefits in the regions where the company operates. Companies can create shared value in three ways: reconceiving products and markets; redefining productivity in the value chain; and building supportive industry clusters at the company's locations.

Although the term "shared value" has only gained widespread use in recent years, the idea behind shared value has been a part of Nestlé's culture from its origins more than a century ago when Pharmacist Henri Nestlé developed and patented the first infant cereal in order to save his neighbor's child from dying of malnutrition. Over the past decade, Nestlé has made shared value an increasingly explicit part of its corporate strategy across its many product lines in all parts of the world. From the 75 million micronutrient fortified Maggi bouillon cubes sold every day in West Africa to the probiotic yogurts specially designed for elderly consumers, Nestlé has shifted its fundamental corporate purpose from merely being a food company to becoming a health, wellness, and nutrition company.

We identified the Moga district as an example of shared value based on our observation of the Moga case.[2] This book is the first in-depth study by a respected independent scholar that seeks to rigorously assess Nestlé's impact on the Moga

---

[1] Michael Porter and Mark Kramer, *Creating Shared Value*, Harvard Business Review, January 2011.

[2] Michael Porter and Mark Kramer, *Strategy and Society,* Harvard Business Review, December 2006.

community. Professor Biswas has documented the extent to which Nestlé has reduced poverty, improved wellbeing, and fostered environmental conservation across the region. In so doing, he has identified even greater social impact than our previous article described. Nestlé's investments in the Moga district have improved farmers' livelihoods and dramatically increased milk production across the region, creating direct employment for over 2,000 people and indirect employment in Nestlé's suppliers for an estimated 86,000 others. At the same time, Nestlé's efforts in Moga have improved its own supply chain, grown production, increased producer quality, and expanded the local market for Nestlé products. The result has been steadily increasing profits for the company and its shareholders.

Nestlé did not undertake their efforts in Moga to bolster its reputation, expand philanthropy, or pursue corporate social responsibility. Instead, drawing on its history of working with smallhold dairy farmers in Switzerland and in many other countries, Nestlé recognized that building a successful dairy business depended on increasing the volume and quality of milk produced by local farmers. The company understood that these farmers—often poor families milking a single undernourished cow—could not increase productivity without expert advice on how to raise healthy animals; access to technology and financing for drilling water wells; the establishment of cooling facilities and distribution systems to collect the milk; building testing centers to ensure high quality; and above all, making a steady commitment to purchase the milk produced at fair prices, week after week, year after year.

This combination of expertise, resources, and opening markets is the essential formula that shared value corporations bring to economic development. Today, Nestlé employs over 1,000 agronomists and 15,000 farm extension workers that work with over 690,000 smallhold farmers around the world to improve their productivity, raise their incomes, conserve water, reduce environmental impacts, and create economic opportunities for women. Nestlé's investments in Moga and other regions of the world exemplify the powerful role business can play in driving social and economic progress. As companies create shared value by addressing social needs such as these, the capabilities and scalability of business is unleashed to meet societal challenges. Public, non-profit, and multilateral organizations remain critical actors in addressing these challenges. But capitalism, guided by the pursuit of shared value, is a decisive partner in delivering sustainable, scalable solutions to the world's most urgent problems.

We commend Nestlé for undertaking this comprehensive study even though, as Professor Biswas acknowledges, the absence of baseline data from decades ago created significant limitations on the study's methodology. Optimizing shared value creation by any company requires setting explicit, measurable business and social objectives at the outset, accurate baseline data, and real time measurement of results. Companies cannot know the extent to which they are creating shared value if they do not measure their progress on social objectives and, importantly, the degree to which their social performance improves economic value for the business.

When companies fail to understand or rigorously track the interdependency between social and business results, they miss important opportunities for innovation, growth, and social impact at scale. Yet the tools to measure shared value in order to understand the link between business performance and social value creation are just beginning to be developed. Even the companies like Nestlé, that are advanced in pursuing shared value, are only beginning to put in place the tools and processes to rigorously measure shared value on a contemporaneous basis in order to optimize their results. This study is an early and important step toward that end. Moving forward, Nestlé can use this information as a baseline to link future progress on social objectives to business results. Doing so will enable Nestlé to enhance shared value even further.

Since the publication of our article, companies from around the world have stepped forward to embrace the idea of *Creating Shared Value* and to rethink their strategies to incorporate it into their core business purpose. Developing the tools to rigorously and objectively measure social impact is an essential component of any such strategy. We encourage all companies to follow the footsteps of leaders like Nestlé and begin rigorously measuring the link between social and economic performance and use these insights to inform and improve their shared value strategies.

Michael Porter
Bishop William Lawrence University Professor
Harvard University
Boston, USA

Mark Kramer
Managing Director, FSG and Senior Fellow
CSR Initiative Kennedy School of Government
Harvard University
Boston, USA

Preface

This book is the result of a study carried out by the Third World Centre for Water Management, Mexico, on the impacts Nestlé in Moga, Punjab, India, has had on the social and economic development in the region. The objective of this study is to learn to what extent Nestlé may have contributed to the community's overall societal aspirations, poverty reduction and alleviation, environmental conservation and general improvements in the people's standard of living and overall welfare.

This assessment is the result of a thorough and intensive visit to the Moga factory and surrounding communities in 2011. During this period, the visiting team comprehensively interviewed the factory's current and retired staff, as well as some 198 farmers, suppliers, and ancillary business partners. Individual and group interviews were conducted in Punjabi, Hindi, and English and were often carried out without the presence of any Nestlé representative so as to create a stress-free setting for the interviewees to voice their opinions freely. Such interviews were carried out in the Moga factory itself, premises of several ancillary companies, milk agencies, individual households and farms, as well as during the 50th Anniversary celebrations of the factory where many first and second generations of farmers and collection agents were present. We also interviewed several of the community's political leaders to get their views on the impacts the company has had in the region. Altogether, 211 persons who were not current employees of the Moga factory were interviewed. This included 13 ex- or retired Nestlé employees.

The team spoke to farmers, collection agents, village elders, and other stakeholders as they performed their daily tasks in order to pose minimal disruptions to their working day and also to gain as broad an exposure as possible to understand and assess the milk collection process. This study reflects the interviewees' recollection of the conditions in Moga when the Nestlé factory first opened; their views on how has the situation evolved over the years; what may have been the factory's contributions and impacts in improving their standard of living and quality of life; and how the towns and the villages may have been affected by the presence of the factory. This process was supplemented by an extensive desk review of primary and secondary data available from the Moga and other Nestlé offices, statistical analyses, and a detailed assessment of the body of literature available on India's dairy sector and the region's overall development. These are listed in a comprehensive bibliography at the end of this issue.

We very much appreciate several discussions with T. S Sandhu and D. S. Gill. Notably, T. S. Sandhu's unpublished memoir, as well as his numerous unpublished documents, proved to be a treasure-trove of information of a bygone era. The insights they both shared with the study team as to how the factory and the region have evolved over the years were of immense benefit. For the sake of history, the authors consider Nestlé should keep the documentation, at least in digital form, so that people realize and appreciate the corporate thinking and performance of the older generations who, by all accounts, were well ahead of their time. This is especially relevant since information on socio-economic indicators of the Moga region even as late as for the 1970s are conspicuous by their absence.

A study like this could not have been carried out without the strong support of the Nestlé staff, both present and past, at the Indian headquarters in Gurgaon, and those at the Moga factory. We are most grateful to Sanjay Khajuria, Senior Vice President, Corporate Affairs at Nestlé India, without whose support and encouragement this study could have never been prepared. We also very much appreciate the help of Smriti Verma, who facilitated many meetings and collection of data, as well as provided us with additional information after our field visit as and when needed. At the Moga factory, we received unstinted assistance from everyone we talked with, especially Martin Roemkens, Factory Manager; Babarjit Singh Bhullar, Senior Manager, Corporate Affairs; S. S. Arora, Milk Development Coordinator; Aman Bajaj, Executive, Corporate Affairs; Arvind Malik, Executive, Agri Services; Satish Bansal, Assistant Manager, Agri Services; Rohit Ravinder, Agri Services, Officer; Naresh Mehndiratta, Executive, Agri Services; and Mary Innocentia, Accountant, Agri Services. Aman Bajaj facilitated the organization of most of the interviews with the farmers and retired or ex-Nestlé staff. The team was rather fortunate to be in Moga in October 2011 when the 50th Anniversary of the factory was celebrated. During that event, the team talked to many of the farmers who provided milk in 1961 when the factory first opened, and several of the next generation of suppliers and agents.

Last, but not at least, we are very grateful to Thania Gómez for her secretarial support.

Asit K. Biswas
Lee Kuan Yew School of Public Policy
NUS, Singapore

Cecilia Tortajada
Third World Centre for Water Management
Mexico

Contents

Acronyms

ADB	Asian Development Bank
AES	Agricultural Extension Services
CO_2	Carbon Dioxide Emissions
CHF	Swiss Francs
CSR	Corporate Social Responsibility
CSV	Creating Shared Value
EU	European Union
FAO	Food and Agriculture Organisation
FLA	Fair Labour Association
FTAS	Farm Technical Advisory Services
GADVAASU	Guru Angad Dev Veterinary and Animal Sciences University
GJ	Gigajoules
Ha	Hectares
IDRA	Industries (Development and Regulation) Act
INR	Indian Rupees
Milkfed	Milk Producer's Federation Limited, Punjab State Cooperative
M	Millions
MMPO	Milk and Milk Products Order
NDDB	National Dairy Development Board
NRI	Non-Resident Indian
PAU	Punjab Agricultural University
UN	United Nations
UNICEF	United Nations Children's Fund
VAT	Value Added Tax
Verka	Punjab State Cooperative Milk Producers Federation
WFP	World Food Programme

Chapter 1
Executive Summary

Can business and society work together to foster each other's prosperity? No company works in a vacuum. Invariably, interdependent interrelationships develop between a business and the community within which it operates. The main consideration should be to maximize the nature and extent of mutually beneficial interrelationships. Henri Nestlé, after whom the company is named, was a German pharmacist. In 1867, he formulated *Farine lactee*, a combination of cow's milk, wheat flour and sugar, to save the life of a neighbour's child. Ever since then, nutrition has been the cornerstone of Nestlé business. More than 145 years later, as Nestlé gradually became a major Fortune 500 multinational company, it has continued to evolve, but always with a deep understanding and appreciation of its roles within the societies in which it operates. This suggests that the importance of social development is deeply ingrained in the company's genes.

Nestlé now has multidimensional motivations and aspirations which include increasing profitability, forging strong relations with its business partners, becoming an attractive employer and drawing the best talents, securing customer satisfaction and loyalty, opening up new business opportunities and larger market shares, enhancing international reputation and attempting continuously to improve the social and the economic environment within which it operates. Over nearly a century and half of its existence, as the company has grown and become global, both its business and policies have steadily evolved with time. It has continuously fine-tuned its corporate practices to support its long-term goals and objectives over time and geographical areas. It has consistently strived to strengthen lasting partnerships with its suppliers, shareholders, employees, customers, surrounding communities and various levels of governments, to whom it has increasingly contributed to their tax base and enhanced local economic resilience.

The present publication is the result of a study carried out by the Third World Centre for Water Management, Mexico, on the impacts Nestlé's activities in Moga, Punjab, India, have had on the social and economic development in the region over the past 50 years. The main objective was to learn to what extent has Nestlé contributed to fulfilling the societal aspirations and expectations of the people working in and around its factory in terms of employment generation, poverty alleviation, general improvements in the community's standards of living and

A. K. Biswas et al., *Creating Shared Value*, SpringerBriefs on Case Studies of Sustainable Development, DOI: 10.1007/978-3-319-01463-0_1, © The Author(s) 2014

environmental conservation. The study also assessed to what extent the company has created shared value for itself, Moga's milk farmers, ancillary firms, and the community at large. To the best of our knowledge, this is the first comprehensive and independent study on the impacts of a private sector company in and around the area where they are located that has ever been carried out anywhere in the world.

The findings indicate that Nestlé has very successfully established long-term positive and symbiotic relations with its primary stakeholders in Moga (namely direct employees, suppliers, communities, and customers) by linking social progress with corporate success and profitability. It has formulated and implemented business strategies that deliver economic, social and environmental returns a private sector company expects whilst spreading the benefits of its commercial activities with those it relies on to thrive. Over the years, it has continued to build the tangible and intangible infrastructure (physical, educational, labour, social, and environment-related) needed to transform the existing conditions to its advantage as well as to those who live in and around the area. It has done so by matching corporate objectives to local opportunities, endowments and needs. This has also resulted in positive spillovers and externalities, benefitting local consumers, producers, the community at large and, surprisingly, to a significant extent even the company's competitors. A body of skilled farmers, experienced milk producers and competitive ancillary firms has contributed to make the area more attractive to new investments from other private and public firms. As new milk and dairy businesses have arrived in the region, they have taken full advantage of the expertise and experience from the readily available trained farmers and skilled workers who are now well-recognized for producing consistently high quality outputs. This would not have happened to the same degree, scope and speed had Nestlé not invested in the capacity development of the region.

In Moga, the company has created both direct and indirect employments, attracted capital (physical, investment and human) to the area, and improved the quality of life of a few hundred thousand families. In the process, it has created substantial wealth for its shareholders. In order to ensure long-lasting and sustainable growth, Nestlé has established, fostered and furthered mutually reinforcing beneficial synergies with its suppliers, employees and the communities they belong to. It has not overlooked the fact that it is them who constitute the company's first tier of consumers and that as societal prosperity improves and their productivity increases, so do the company's financial position and competitiveness.

As a socially responsible, responsive and engaged company, it has gone out of its way to consistently support the community both through its business practices as well as through other social initiatives addressing specific needs of the communities of the region. This integration of business practices with the needs of the local communities has secured the human, physical, social, technological and political capital it has needed for operating and growing as a profitable enterprise. All these have reinforced Nestlé's market share, overall reputation and credibility not only in the region and in the country but also around the world. In turn, this has further accelerated its turnover and profits, which has pleased its shareholders and has consolidated their support for the company.

Terms denoting the positive relationship that should exist between the private sector and the society, most commonly known as 'Corporate Social Responsibility' (CSR), are now in the lexicon of nearly all important multinational corporations, numerous academics, non-governmental organizations, consumers associations and also in the list of priorities of many business leaders. Yet, when the Moga factory was first established in 1961, companies very seldom thought about improving the social and environmental impacts of their commercial activities. Neither a concept capturing this thought nor the philosophy behind it had yet been coined or formulated. Moreover, at that time it was almost unthinkable to tie a company's economic efficiency to the social progress of the community where it was established. This is not surprising since knowledge, mindsets, business, and societal perceptions and expectations regarding the role of the private sector were very different half-century ago compared to what they are at present.

Yet, 50 years ago, Nestlé positioned itself at the forefront of the dairy industry in Moga, building and expanding the local market for quality milk products. It did so by and through meeting societal needs and deliberately devising social and economic advancements for the thousands of small farmers in this very impoverished region of Punjab. The philosophy which Nestlé single-handedly developed later, 'Creating Shared Value' (CSV), requires that business practice is firmly grounded on a 'community of interests' between farmers, factory workers, ancillary companies and local consumers. It is a corporate model that goes well beyond the current generic, remedial and damage control philanthropic activities to which the term Corporate Social Responsibility (CSR) is usually applied.

It is thus impressive that even though the concept of Creating Shared Value did not exist five decades ago, the general philosophy behind it was implicitly, and even somewhat explicitly, evident in Nestlé's actions when it decided to build a factory in Moga and establish a milk district there. In retrospect, by reviewing and assessing the approaches the company took and has taken at Moga, one can only applaud at the implemented activities and at the extent to which the company has tied social progress to corporate success. These actions categorically show that even though the unknown and unsung initial planners of the Moga factory did not explicitly articulate the thinking behind Creating Shared Value, or even Corporate Social Responsibility, they fully appreciated the importance of such a business model and philosophy. They also successfully implemented them in their business plans.

It was not until 2006 that Nestlé became the first multinational company in spelling out, streamlining and adopting the Creating Shared Value approach. In Moga, many decades before that, those behind the planning of the factory were quick to realize that social improvement and progress did not thwart corporate growth and profitability. In fact, the success of the company and community welfare are mutually reinforcing processes. They were convinced that the success or failure of the Moga factory was intertwined with the betterment of the lifestyles of its many milk suppliers, the wide array of entrepreneurs, labourers providing other inputs and the local staff of the factory itself. They realized that if the people's standards of living could be improved, the success of the factory would be assured, and this would create a virtuous and mutually beneficial cycle.

A very important factor has also been the interest of the company in developing long-term relational interactions, and not only transactional ones, with the myriad of small-scale producers it has worked with. This has been a fundamental instrument to build trust, confidence and loyalty with its suppliers, consumers, and the rest of the society they are conjoined with. In fact, and since the inception of the factory in the early 1960s, the community and the various levels of government have consistently regarded the Moga factory as a positive 'agent of change,' and also as an 'engine for rural development'. They have consistently acknowledged the advantageous externalities, beneficial spillovers and favourable multiplier effects it has brought about in the region. All of these positive impacts have directly and indirectly resulted in a substantial and lasting change in the population's livelihood, overall welfare and steady improvements in quality of life.

Moga has undergone changes well beyond significant improvements in the community's standard of living, social and educational conditions, food habits, hygienic practices and infrastructural developments. For many years, the lasting changes in livelihoods, assured and regular income receipts, steadily improving socio-economic conditions, the building of intra- and inter-company transferable capacities and skills, and loyalty stood at the core of the suppliers' decision to work continuously with Nestlé. Now, however, our findings indicate this situation is changing as local competition for the procurement of milk and other associated raw materials intensifies. Farmers and suppliers are becoming increasingly aware of the high demand for their products and skills from other sources and players that did not exist before. As the farmers' bargaining power increases, firms seeking to seize a share of the local market need to respond to the producers' demands in terms of more competitive terms of engagement.

This poses Nestlé with both a challenge and an opportunity to reassess its business model, respond to present needs and meet future challenges. Undoubtedly, this could also be an occasion for the company to become, once again, a pioneer in successfully linking and mainstreaming corporate success to community development and social wellbeing. Concurrently, it could remain as a competitive local, regional and global player under an ever evolving and challenging panorama where change stands as the only assured constant. An overall view of what has been the interplay of factors affecting Moga's development since Nestlé established its factory in the area is shown in tabular form in Annex 1. How the various factors have changed from 1961 when the factory was first established to 1981 and then to 2011 are indicated in Annex 2.

Chapter 2
Introduction

Nestlé has just completed one centenary of its involvement in India, where it started its trading activities in 1912. During this long 100 year partnership, both the company and the country have evolved very significantly. Some 50 years ago, very few people, if any, could have predicted that India would jettison its image of 'Licence Raj' and become an emerging economic world power. Neither would have they been able to foresee that Nestlé would be the world's biggest company in the food, beverages and nutrition sectors and a trailblazer in fostering the interdependence between commercial activities and social goals, now widely embraced by the corporate sector as Creating Shared Value. This enduring, growing and symbiotic partnership between the company and the country has served both parties well as they continue to reap long-term, mutually reinforcing and rewarding benefits in every sense. It also stands as a good example of how private sector participation can seize local opportunities to trigger rural development whilst supporting national priorities and goals.

The decade following India's independence in 1947 was a socially and economic difficult period for the country. For nearly the subsequent five decades, India followed an import substitution industrialization model as the overriding socio-economic philosophy to stimulate domestic manufacture and contribute to self-sufficiency. The country's *dirigiste* approach was guided only by the belief that "India should produce whatever it can and India should export whatever it produces" (Bhagwati and Desai 1970: 466). The philosophy behind the model was to reduce foreign dependency by locally manufacturing necessary products and creating an internal market to sell them (Rodrik 2005).

The Government thus formulated and imposed an overall framework under which all industrial development activities would occur. The 1951 Industries Development and Regulation (IDRA) Act laid the foundations for administrative control on all manufacture and production. Simultaneously, the centrally planned economy imposed a series of controls on inputs, outputs, land, labour, etc. Prices, quantities, quality and production targets were all State-determined and supported by a highly convoluted and officious system. Over time, licensing requirements became increasingly complex and stringent and were subjected to arbitrary changes in Government policies and bureaucratic dictates. Industrial activities were

A. K. Biswas et al., *Creating Shared Value*, SpringerBriefs on Case Studies of Sustainable Development, DOI: 10.1007/978-3-319-01463-0_2, © The Author(s) 2014

subject to a myriad of procedures, often somewhat arbitrary, and requiring the approval and clearance of overlapping and uncoordinated ministries.

Nestlé had been exporting condensed milk and baby food to India since 1912. However, in 1958, the Indian Government decided to ban imports of milk powder and other manufactured milk products to conserve foreign exchange and stimulate domestic production capacity. Within this State-planning system and inward-looking industrialization policies, Nestlé was invited to consider the possibility of establishing a milk district and setting up a dairy factory in Punjab. In 1959, intensive negotiations between the Indian Government and Nestlé took place to define the area from which the company could be permitted to buy fresh milk. The area was finally fixed at 11,000 km^2 around the village of Moga in the state of Punjab.

At the time of the construction of the factory at Moga, the area was a place of abject poverty, widespread malnutrition, mud-built houses and low-productivity. Subsistence agriculture was the main economic activity. Over 80 % of the population lived in small villages averaging 250 resident families, with six persons per household. Water supply, irrigation systems and transportation were mostly animal-operated. Houses did not have access to piped water. Sanitation facilities were primitive and only very few scattered villages had access to electricity (Sandhu 1978, 1981a, b, c, d; 1988; Nestlé 1993). The area's challenging socio-economic conditions might have been one of the reasons as to why the Indian Government selected Moga for Nestlé to establish a factory there, hoping its presence would contribute to its social and economic development and the welfare of the population within a reasonable timeframe. However, this is only a hypothesis. Despite intensive research, no historical document could be found as to why the Indian Government selected Moga.

Before production could start, Nestlé had to secure the supply of its main input, milk, in an area where there was no milk culture and the dairy sector was largely informal and unregulated. Milk sold was subject to widespread adulteration and arbitrary marketing practices. Not only was dairying a low-input, low-output supplementary income-generation activity to agriculture, taboos and social-religious reservations regarding the commercialization of milk discouraged sustaining, let alone increasing production levels.

Despite the country's long historical tradition of dairying, poor breed, neglect of animal health and lack of proper and adequate feed, ensured that the then prevailing milk yields were extremely low. Moreover, the region almost had no dairy cows and whatever limited buffalo milk was produced was used primarily for household consumption. The milk culture and knowledge and expertise on milk production simply were not available in the area. Faced with these serious constraints, and honouring the agreement made with the Government of India, Nestlé planned to take all the necessary steps to transform the area into a thriving milkshed.

In addition, all too often changing import substitution policies posed significant problems to the private sector as the Government unilaterally altered industrial policies without any serious discussion or consultation with the involved stakeholders. Policies on dairy development were no exception. Not only were many business operations micromanaged by numerous Government dictates, but also the companies had to permanently look out for sudden, and imposed requirements that could substantially alter the sustainability and the overall economics of their prevailing business models and thus profitability. A good example of this sudden shift came in 1972, when the milkshed area was re-demarcated and part of the area where Nestlé was allowed initially to collect milk was ceded to various other Government milk schemes (Food Specialities Limited 1972).

By 1991, when the 'Licence Raj' model was being dismantled, Nestlé's exclusivity to buy milk within a specified area was completely withdrawn and competition for buying milk from the area intensified because of the arrival of new competitors. Following national economic reforms and liberalization, the Milk and Milk Products Order (MMPO) came out in 1992. This legislation was put in place to regulate the local production of milk and dairy and included sanitary and hygienic controls to ensure product quality.[1]

By then, and after sizeable investments and efforts, the company had successfully developed and organized a reliable and sustained dairying culture and a functional supply chain in the area. The challenges Nestlé faced during the first three decades, first to successfully establish a factory and then to assure its steady expansion, were truly complex as well as immense. They ranged from rigid input and output pricing policies, to poor infrastructure and the total absence of a dependable and long-term framework for promoting a competitive dairying sector.

References

Bhagwati JN, Desai P (1970) India: planning for industrialisation. Oxford University Press, London

Food Specialities Limited (1972) Report and accounts for the year ended 31st December 1971. New Delhi, Directors' and Auditors' Report

Karmakar KG, Banerjee GD (2006) Opportunities and challenges in the Indian dairy industry. Tech Dig 9:24–27. Retrieved 17 June 2011 from http://www.nabard.org/fileupload/DataBank/TechnicalDigest/ContentEnglish/issue9td-6.pdf

Nestlé (1993) Nestlé in India 1962–1992. Nestlé, Vevey

Rodrik D (2005) Growth Strategies. Handbook of Economic Growth 1:967–1014

Sandhu TS (1978) Agro-climatic and other situations. Year-book. Food Specialities Limited, Moga

Sandhu TS (1981a) Extent of full employment generated by Moga milk plants in its milk shed area. Food Specialities Limited, Moga

[1] India's compulsory regulatory legislation for the dairy industry was long framed in two main documents. The Prevention of Food Adulteration Act, 1964, includes a section specifying the standards and minimum quality levels to be achieved in milk production. Additionally, the 1992 Milk and Milk Products Order requires all units handling more than 10,000 L of liquid milk per day or more than 500 tonnes (t) per annum to get a processing and distribution permit. The order also specifies the milkshed area and lists the basic hygienic conditions to be maintained at the premises (India's TEDO, undated). In 2011, the Food Safety and Standards Authority issued an overarching regulation to consolidate numerous previous policies regarding food products including milk and milk products (Karmakar and Banerjee 2006).

Sandhu TS (1981b) Evaluation of development activities of Moga milk plant. Food Specialities Limited, Moga
Sandhu TS (1981c) Inputs for milk procurement. Food Specialities Limited, Moga
Sandhu TS (1981d) Prosperity? Food Specialities Limited, Moga

Chapter 3
Moga Factory: The Beginning

After some background fact-finding and analysis of the situation, Nestlé decided to take the risk and submitted a plan for the construction of a milk factory in Moga, some 160 km northwest of Chandigarh, the capital of the state of Punjab, for approval by the Government of India. The formalities pertaining to the incorporation of an Indian company were successfully completed in 1959, in which Nestlé owned 90 % of the capital. Accordingly, in 1961, in a non-descript and a very little known area of Punjab, with no culture of sustained milk production and a virtually non-existent milk economy, Nestlé constructed a factory with a processing capacity of 9,000 t of fresh milk per year. Moga also became the first registered office of the company in India.

The site for the factory was selected by Nestlé on the important criteria of its proximity to the railways so that inputs and products could be easily transported. After the site was identified, land was purchased in three different lots. Even though the factory was close to the railway station, transportation of machinery and other goods was difficult since every piece had to be transported by bullock carts, which at that time was the primary means of transportation in the region.

Whilst socio-economic and demographic changes, rising income levels, rapid urbanization and changing food habits and lifestyles guaranteed a growing domestic market for manufactured milk products, especially towards the last two decades of the 20th century, the regular availability of enough milk was a serious concern. When the factory opened, the most pressing challenge Nestlé faced was whether the farmers of the area could produce 9,000 t of excess milk, and whether the company could purchase and process it as the basic raw material to produce appropriate dairy and other manufactured products.

In light of the considerable challenges ahead, right from the very beginning Nestlé made some important policy decisions to guarantee the successful creation of a milk district in Moga. In retrospect, such measures proved to be not only correct but also turned out to be more appreciated a half a century later. For instance, the company determined it would not develop its own dairy farm, nor would it own any milk cows. Instead, it would primarily buy milk directly from the farmers of Moga and reduce the intervention of the middlemen who often made most of the profits from these transactions. It would buy the milk available for sale

A. K. Biswas et al., *Creating Shared Value*, SpringerBriefs on Case Studies of Sustainable Development, DOI: 10.1007/978-3-319-01463-0_3, © The Author(s) 2014

from all the villages within the allocated area and from all types of producers. It ranged from landless labourers who might have one, or at most two, female buffaloes and thus could sell only few litres of milk occasionally to the company, to large farmers who could supply reasonable quantities of milk on a regular basis.

Aware of the local economics, lifestyle, conditions and opportunities, the company decided that it would not encourage the farmers to give up their already established agricultural practices of growing wheat, maize or cotton and focus primarily on milk production. Rather, it would encourage them to consider milk as a source of additional and regular income throughout the year, very much like an additional 'cash crop'. This approach would enable the farmers to meet much of their day-to-day financial needs from the formal milk sales to the company. Moreover, milk could be produced without jeopardizing their then agricultural activities. When the cropping season was over, farmers could store some crops for their use until the next season and thereafter sell the surplus for additional income, and store some as cattle-feed. In retrospect, the policy proved to be a 'win–win' situation for both the farmers and Nestlé on short-, medium- and long-term basis. These measures helped integrate the company into the local community and economy as well as to locate its operations close to the milk-producing farmers.

The milk collection process for the Moga factory started on November 15, 1961. On the very first day, only 180 farmers from four villages supplied a total of 511 kg of milk, an average of 2.84 kg. During the first full calendar year of collection, in 1962, only about 2,054 t of milk could be procured directly from the farmers. This was less than one-quarter of the plant capacity. Thereafter, and as the company's investments in farmers' technical capacities, capital, and standard of living began to pay off, milk procurement increased steadily and substantially. By 1970, it was evident that Nestlé needed significant additional investments for new buildings and further infrastructure to increase the plant's processing capacity in order to keep up with the high rates of milk procurement. This also reflected in the many more and extensive milk routes and collection centres that had to be established to cope with many more farmers supplying milk to the company and thus the procured volumes.

3.1 Challenges and Opportunities

For Nestlé to become an 'agent of change', the company had to effectively and simultaneously overcome a gamut of social, cultural, economic, infrastructural and other attitudinal challenges. Only to start production, the company had to successfully reshape social, religious and cultural attitudes to dairying; organize the milk trade whilst addressing serious infrastructural and production constraints; establish a context-tailored milk collection system; set strict and transparent measuring standards in terms of both quality and quantity of milk purchased; and pay its supplying farmers competitive and appropriate rates on a timely basis which were fair to both the parties. Figure 3.1 shows diagrammatically some of the

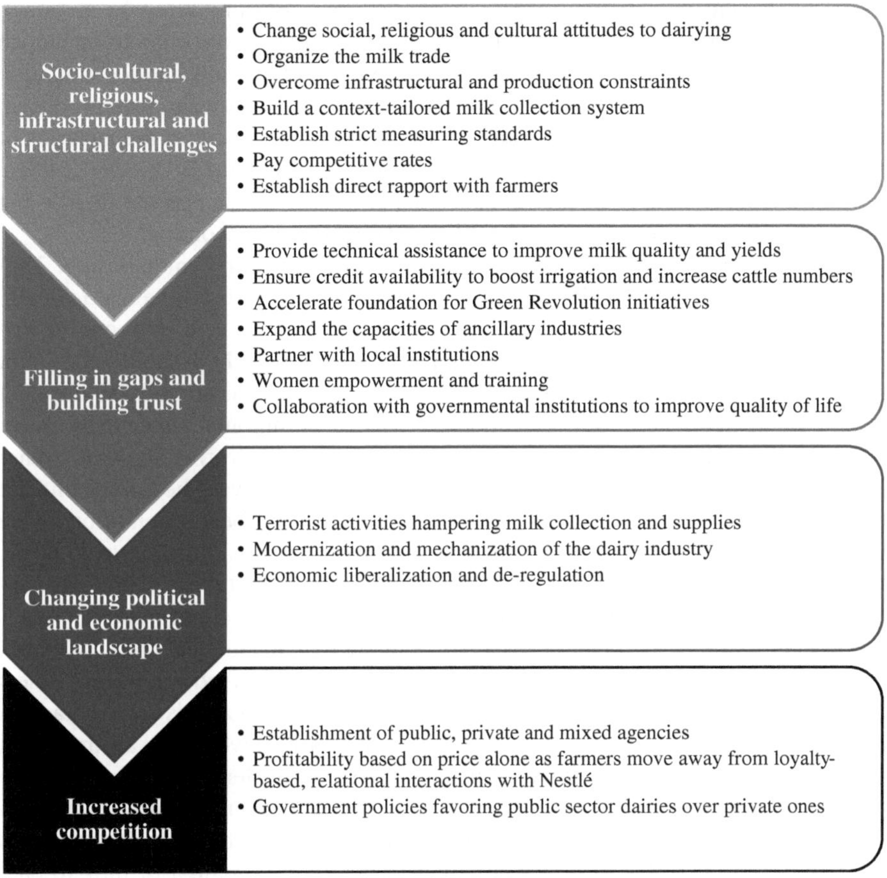

Fig. 3.1 Seizing opportunities, overcoming constraints

different opportunities and challenges determining the local landscape and Nestlé's operations in Moga during the first three decades.

When Nestlé arrived in Moga, the area's agricultural profile consisted of small farms (average 5 acres or 2 ha) where wheat was the main crop. Transportation networks within and between the villages were poorly developed, and for most practical purposes non-existent. They were invariably narrow earthen roads that often became impassable during each monsoon season. Very few villages had access to electricity and irrigation practices had not changed for generations. Only a small percentage of farms near the canals could be irrigated, and canals could carry water for only about 1/3rd of the time (Nestlé 1975; Sandhu 1978, 1980).

Some farmers did have wells but very few of them, if any, could afford diesel pumps. The cost of installing tube wells was around INR 4,000, in 1962, which was equivalent to the price of 10 t of wheat. For average farmers, irrigation was beyond reach because of the pump-related costs. Water was mostly withdrawn from the wells by Persian wheels operated by a pair of oxen or sometimes by a camel. Chemical fertilizers were basically unknown in the area. In the climate of Punjab where irrigation is essential to ensure good crop production, only rainfed irrigation could be practiced in most cases. Thus, only one crop per year was produced. Agricultural production, the mainstay of the farmers, depended on the vagaries of rainfall, varying significantly from one year to another (Sandhu 1978, 1980, 1981a, b, c, d). Thus, economically, the farmers led a precarious existence and their living standards reflected food, housing, health and education shortcomings.

Equally important were the conditions under which dairy animals were kept which posed significant challenges to milk production in the area as well as to health conditions of the farmers. The vast majority of the farmers had only one or two female buffaloes, which produced enough milk to meet the household needs. Not surprisingly, milk yields were low and the lactation period was around 150 days per year. Furthermore, buffaloes were poorly fed and commonly affected by many diseases. Over 60 % of the calves born were infested with worms and many other types of parasites as a result of which they often did not survive. The farmers did not feel that proper feeding of the buffaloes and their overall health conditions were important issues as long as they produced enough milk to satisfy their own domestic requirements (Sandhu undated, 1978,1980). This attitude to dairy animals was closely related to social attitudes regarding the commercial sale of milk.

Poor agricultural output often meant poor animal health and thus low milk yields, which were almost totally for domestic consumption. And yet, when farmers did have a milk surplus, they did not commercialize it openly as this activity faced a major cultural problem. Selling milk to the *dodi* (milk buyer) was an important social taboo for the farmers. It was considered to be equivalent to selling their own sons. From time to time, a few farmers sold to the milk buyer few litres of milk on the sly, when no one was looking, to avoid being ostracized by the neighbours. As such, the *dodi* had to travel from one village to another searching for milk to buy.

Not only selling milk was frowned upon by the society in the area, but also the *dodi* was not known for his honesty. Sellers were often cheated with inaccurate weights, given inappropriate prices for the milk and payments were neither regular nor prompt. In addition, adulteration of milk by the farmers as well as the *dodi* was quite common. Thus, buying and selling milk had an 'unpleasant odour' around it, and this behaviour was generally not favoured by the community at large in and around Moga. If discontinued and irregular milk transactions in this informal sector were so severely socially penalized and discouraged, the very thought of formal and structurally engaging with an agent to sell milk was considered to be unthinkable for most farmers.

To further complicate the situation, villagers were mostly traditional, conservative and, because of their poor economic conditions, risk-averse. There were also a series of cultural, social and economic factors, preconceptions and taboos that made farmers reluctant to make changes to certain aspects of their regular economic, public and private activities. For example, consumption of *deshi* (native) liquor was quite common and treatments of maladies by *tona* (magical healing by shaman) were widespread.

There were several technical challenges that had to be overcome as well. For example, in the area that was allocated to Nestlé, dairy farming with cows was basically unknown. In fact, there was no history in India of successfully and economically producing dairy products, not even from more common buffalo milk. And yet, when the plant was set to open, one of the first products it was to manufacture was MILKMAID, which is a sweetened condensed milk, most of which was to be supplied to the Indian army. The company was thus faced with the key issue of whether condensed milk could be successfully produced from buffalo milk, which has a fat content of nearly 2½ to 3 times that of cows' milk.

Nestlé was fully aware of the challenges it faced, but it also knew that the land was fertile and the Punjabis were well known for being dependable and hard-working farmers. They may have been rooted in traditions but they could be persuaded to adopt new agricultural practices, especially if they could be convinced that these would contribute to improving their own financial situations and the standard of living of their families. Along other propitious endogenous factors, it was this attitudinal openness to technological change and novel techniques and ideas that turned Punjab into India's 'bread basket', the epicentre of the Green Revolution, and a successful example of the development of the dairy industry (Agarwal 1983; Bhalla 1983; Chaudri and Dasgupta 1985).

References

Agarwal B (1983) Mechanization in Indian agriculture: an analytical study based on the Punjab. Delhi School of Economics, New Delhi

Bhalla GS (1983) Green revolution and the small peasant: a study of income distribution among Punjab cultivators. Concept Pub. Co., New Delhi

Chaudhri DP, Dasgupta AK (1985) Agriculture and the development process: a study of Punjab. Croom Helm, London

Nestlé Alimentana Company (1975) Nestlé in the developing counties. Nestlé, Vevey

Sandhu TS (1978) Agro-climatic and other situations. Year-book. Food Specialities Limited, Moga

Sandhu TS (1981a) Extent of full employment generated by Moga milk plants in its milk shed area. Food Specialities Limited, Moga

Sandhu TS (1981b) Evaluation of development activities of Moga milk plant. Food Specialities Limited, Moga

Sandhu TS (1981c) Inputs for Milk procurement. Food Specialities Limited, Moga

Sandhu TS (1981d) Prosperity?. Food Specialities Limited, Moga

Sandhu TS, Dhaliwal KS (1980) Rural Poverty. Food Specialities Limited, Moga

Chapter 4
India's Revolutions: Seizing Local Potential and Harnessing National Opportunities

4.1 The Green Revolution

In retrospect, the construction of the factory at Moga could not have been any timelier. The Green Revolution reached this region of Punjab around 1965–1966, shortly after its opening, when the farmers were encouraged by the Government to grow hybrid wheat and corn, which produced much higher yields. This intensification of agricultural production resulted into higher output even when land area cultivated augmented only marginally, and more importantly, it brought about higher earnings and many socio-economic advantages for many rural communities. Encouraged by such positive developments, they were more willing to experiment with new ideas which could further improve their standards of living.

In 1968, as infrastructure improved, new high-yielding seeds were introduced and agriculture was mechanized, wheat yields increased by 50 % in the area. This meant an additional income of INR 400 per acre. Better irrigation and improved access to agricultural inputs and capital also allowed many farmers to start growing rice from July to October. Soon after, rice production saw a 2.5 fold increase (Food Specialities Limited 1991; Nestlé 1993). This further ensured that the development of the milk sector and its introduction as part of the domestic economy did not come at the expense of agricultural activities or outputs. Instead, it gave a complementary boost to the region's agricultural economy and family incomes.

The Green Revolution brought about significant improvements in agricultural intensification and thus was compounded by the benefits from farm economics that was boosted by the establishment of the Moga factory in the area. Concurrently, a prosperous community triggered the inflow of migrant workers from Bihar and Uttar Pradesh. Their synergistic interactions allowed Moga's farmers and even labourers from other states to provide their children and other family members with education, better food and nutrition and improved health services. These new and improved non-farm employment opportunities were also seized by women and young farmers of the region. These played an important role in slowing down rural outmigration.

Between the incomes generated from milk production and the economic progress of the Green Revolution, the quality of life and the standards of living of the

A. K. Biswas et al., *Creating Shared Value*, SpringerBriefs on Case Studies of Sustainable Development, DOI: 10.1007/978-3-319-01463-0_4, © The Author(s) 2014

people in and around Moga went through a radical improvement that proved to be unprecedented in the local history (Akram-Lodhi 2008). But this was not to be the only radical change the region was going to experience.

4.2 Operation Flood and the White Revolution

Milk is a cultural, religious and socially important food in Indian culture and as such has an important place in the Indian diet. For the country's many vegetarians, it constitutes the only acceptable source of animal protein and for the most impoverished sectors of society it is a relatively cheap and nutritious product (Karmakar and Banerjee 2006).

Seeking to achieve self-sufficiency in this important foodstuff, the Government of India engaged in a nation-wide scheme that aimed at sustaining rural livelihood and link organized milk-producing farmers in rural areas with urban consumers. In 1970, the National Dairy Development Board (NDDB) started the First Phase of 'Operation Flood', a programme to commercialize, mechanize and formalize the production, processing and marketing of milk along cooperative lines. The programme was initially financed by selling the powdered milk and butter oil donated by the European Economic Community (now European Union) through the World Food Programme (WFP) as a form of food aid. Operation Flood encouraged the establishment of village milk producers' cooperatives and helped change the rural economic landscape in many small communities across India.

The White Revolution, as this initiative was subsequently known, has now received widespread national and international recognition and numerous awards (FAO 2005; Singa 2007). Similarly to other initiatives surrounding livestock development and the consequent improvement in the welfare of the country's rural population, Operation Flood is also credited with having lifted many millions out of poverty. This help raised the profile of the milk sector and drew attention to this promising rural development initiative. These achievements of the public sector are widely known in India and abroad but very few are aware of the pioneering efforts that Nestlé undertook in Moga.

The study team was of course fully familiar with the contributions of the White Revolution to India's rural and urban sectors. It was also prompt to identify many similarities in the procurement model concept that Nestlé had developed on its own to create a milk market in Punjab which did not exist before. Any objective observer analyzing the results and impacts of the Nestlé activities in Moga on rural development and poverty alleviation would have to conclude that the company efforts have been as successful as Operation Flood in catalyzing the multi-purpose role the dairy sector plays in rural economies. When Nestlé started to organize the farmers of Moga in 1961, it created a milk market and successfully formulated a business model that has worked very well under the prevailing Indian conditions. The fundamental question that arises is if Nestlé created a very successful milk procurement model similar to Operation Flood and its impacts on rural development and urban consumers were similar to those of Operation Flood, why very few people even in India are aware of its contributions? Although the team did not have the time to analyze and reflect on this anomaly, few reasons can be put forward.

First, India, ever since its independence, has consistently favoured public sector over private sector not only in terms of overall philosophy but also in terms of policies. Only a few of these public policies and institutions have been as successful as Operation Flood. Not surprisingly, Central and state governments have spent considerable time, energy and funds in publicizing the justifiable success of this public sector effort. This is understandable given the relevance of the dairy sector in India's rural areas and the potential contributions to rural development any initiative fostering livestock activities may bring about. More than 30 million small producers are involved in milk production alone and there are around 80 million farm households engaged in livestock raising currently reaping benefits from enhanced nutrition, additional income sources and productive employment, mostly for women (FAO 2005; World Bank 1999).

Second, because of the support of national, state and international governmental organizations like the World Bank, Food and Agriculture Organization (FAO) and the Asian Development Bank (ADB), Operation Flood has received very wide national and international publicity and justified recognition globally. To start with, the World Bank loaned funds to Operation Flood, which became one of its most successful projects. Thus, not surprisingly, the Bank has contributed to further promote one of its most successful projects, both nationally and internationally. Consequently, numerous books, papers and monographs have been written on this success story and countless workshops have been organized in different parts of the world where its impacts and the lessons that could be learnt and its potential replicability in other countries could be discussed. These have further publicized its success.

Third, Nestlé has not promoted the success of the activities at Moga as an important development-related case study from which many positive lessons could be learnt, including poverty alleviation, rural development, employment generation, resource conservation and women's empowerment. A quick search in Google identified 116,000 million references to Operation Flood but only 47,100 listing to Moga. Not one of these listings on Moga is a serious and comprehensive analysis of the lessons that could be learnt from this impressive case. In fact, to the best of our knowledge, this study is the first serious attempt to analyze and document comprehensively what have been the impacts of Nestlé's activities on the rural and the urban development of the region.

It is now up to Nestlé to showcase its pioneering business model in this community in Punjab, which as a corporate venture has been a remarkable success. It has made significant contributions to the region's rural development and poverty alleviation efforts. Such initiatives are generally the responsibility of the public sector to perform and are left to the Government to carry them out. In this region, nevertheless, the company has made a deliberate and conscious effort to foster a healthy and productive community of diary villages which has steadily improved their social, economic and environmental conditions.

It is worth noting that a private sector company undertook measures normally left to various Government departments. In the process, Nestlé built a highly successfully business which ensured that all its consumers have access to high-quality milk and other associated food products. It is thus a remarkable case of corporate success, a pioneering experience of Creating Shared Value and an excellent example of corporate social responsibility. What is also worth stressing and capitalizing on is that all these efforts started and were achieved nearly half-a-century ago, when the social demands, theories and corporate visions behind Creating Shared Value and Corporate Social Responsibility were not even enunciated let alone implemented. The company itself, the dairy sector, competitors and the private sector alike have all benefitted from these practices, lessons and corporate culture.

In retrospect, it should be noted that Nestlé developed a successful milk sector and economy in India, when the country was in the early stages of exploring such models. The effects this contribution had in terms of poverty alleviation, capacity building, productivity gains in the dairy sector and overall rural and human development should be included as an important element of India's account of the advances made in the dairy sector. This study shows a useful and informative case of the positive role Nestlé, a private sector company, has played in establishing, consolidating and fostering a cluster culture for milk production, marketing and processing in this region of Punjab.

References

Akram-Lodhi AH (2008) (Re) imagining agrarian relations? The world development report 2008. Agriculture for development. Dev Change 39(6):1145–1161

Food and Agriculture Organization (FAO) (1976) Detailed report of the second world food program evaluation mission on operation floor, Indian dairy corporation. The Economic Scene, No. 7. 10–19

Food and Agriculture Organization (FAO) (2005) Breaking ground: gender and food security. Food and Agriculture Organization, Rome

Karmakar KG, Banerjee GD (2006) Opportunities and challenges in the Indian dairy industry. Tech Dig 9:24–27. Retrieved 17 June 2011 from http://www.nabard.org/fileupload/DataBank/TechnicalDigest/ContentEnglish/issue9td-6.pdf

Nestlé (1993) Nestlé in India 1962–1992. Nestlé, Vevey

Singa OP (2007) Agro-industries characterization and appraisal: dairy in India. Dairy Farmers' Organization, Anand, India. Working paper No. 21. Food and Agricultural Organization, Rome

World Bank (1999) India livestock sector review: enhancing growth and development. Allied Publishers, Mumbai

Chapter 5
Milk Production

5.1 Milk Procurement

Only 10 years after the factory was built, Moga's story became significantly different to that of the national dairy sector. Up until 1970, India's annual milk production grew at a very conservative rate of about 1 % per annum (Delgado et al. 2003; Sharma et al. 2002). In contrast, it can be inferred that the rate must have been considerably higher in the new Moga milk district judging by the almost 13-fold increase in milk procurements that Nestlé witnessed between 1962 and 1972. Such rise also denotes the radical socio-cultural attitudinal change the company triggered in the region's farmers and the community at large to selling milk in the open, formally and regularly. Only a few years after the Moga factory was opened, disapproval in milk commercialization had totally disappeared. Nestlé had gradually built confidence in the milk trade until it was fully integrated in the social, cultural and economic life of the community.

Table 5.1 shows the evolution of fresh milk procurements during the first three decades of the Moga factory. This table gives an indication of steady increases in milk production and thus availability and procurement in the milkshed area allocated to Nestlé. During the first full year of milk procurement, in 1962, the total amount purchased was 2.3 million kg. In only 10 years, the milk culture was established so soundly that the total purchase grew over 10 times. Production has steadily increased over the past decades. Higher milk volumes, from a growing number of producers, invariably translated into higher incomes and better standard of living of an increasing number of farmers.

It is important to note that the main impact was on small and marginal farmers. This group constituted the largest percentage of the people supplying milk to Nestlé and thus benefitted the most from the company's presence at Moga. For example, in 1984, it was estimated that the average income from milk for marginal farmers was about 2.8 times that from the crops they produced, and for small farmers the milk income was a little less than what they received from the crops. Thus, in terms of poverty alleviation, the new and regular source of market income

A. K. Biswas et al., *Creating Shared Value*, SpringerBriefs on Case Studies of Sustainable Development, DOI: 10.1007/978-3-319-01463-0_5, © The Author(s) 2014

Table 5.1 Evolution of milk procurements, 1962–1992

	1962	1972	1982	1992	2012
Direct suppliers	4,460	18,413	34,238	46,308	62,183
Milk routes	3	7	13	20	31
Village dairies	63	230	421	650	1,483
Milk purchased (t)	2,054	26,660	42,050	89,872	236,472
% of total purchase	100 %	100 %	73 %	85.5 %	99.59 %
Farmer's cooperatives	–	–	40	15	–
Milk purchased (t)	–	–	2,380	841	–
% of total purchase	–	–	4 %	0.8 %	–
Indirect purchases[a]					–
Milk purchased (t)	–	–	13,260	14.401	979
% of total purchase	–	–	23 %	13.7 %	0.41 %
Total purchased t	2,054	26,660	57,690	105,114	237,451

[a] Indirect purchases were from freelance milk agents who collected milk from villages that were not covered by milk routes. *Source* Nestlé records

from milk that came from Nestlé was significantly advantageous for the marginal and the small farmers. These groups of farmers constituted some 60 % of the area's population at the very beginning when the company started to operate in Moga. In addition to reducing household poverty, additional income has also been crucial for members of this group to elevate their social, cultural, economic and political status and standing. These important developments and the significant gains from milk sales accrued to marginal and small farmers are shown in Table 5.2. Such major contributions to rural development and economic betterment should not be understated and/or unappreciated as the majority of the poor in the country and in Punjab reside in non-urban areas.

The results of this analysis indicate that there is no question that the presence of Nestlé at Moga has had a very major impact, both economically and socially, on the most vulnerable sections of the society: landless labourers with one or two buffaloes, marginal farmers with farm sizes of 2 acres (0.8 ha) and small farmers with farm sizes of 2–5 acres (0.8–2.0 ha). From a survey that was carried out from November 1, 1991 to October 31, 1992, these three economically lower categories

Table 5.2 Average annual income from crops and milk for 1984

Farmers		Average annual income (INR)	
Type	Percentage	Crops	Milk
Marginal	19.3	3,500	9,800
Small	39.4	13,000	11,000
Large	28.4	25,000	12,000
Very large	12.9	45,000	5,800
Total	100	86,500	38,600

Note Estimated by T.S. Sandhu, September 1985. Marginal farmers were those who owned up to 2 acres of land and had one buffaloes; small farmers were those holding up to 5 acres of land and two buffaloes constituted 78.3 % of Nestlé's milk suppliers and contributed to 43.2 % of its total milk procurement (Nestlé India Limited 1991). For these groups in particular, family and household members are usually in charge of performing all required farm tasks, providing the totality of the needed labour and seldom employing hired workers. As such, any income increases are fully used by the households.

Moreover, the considerable percentage of milk that is procured from small, marginal landless farmers and other workers in rural areas points at the large extent to which these groups have engaged in and benefited from commercial dairy activities. These generate remunerated employment and occupational activities for family members, many of them women, and thus contribute to better living conditions. For example, for marginal milk producers, who are amongst the most economically disadvantaged sectors of society, milk receipts are of great benefit as there are no caps on the amount of milk that they can sell. They rely on Nestlé as a guaranteed buyer for their produce and see the factory as a reliable and integral part of their livelihood. More generally, albeit depending on the season (and thus fluctuating farm earnings), milk-related income is used for feed and fodders; improved farm equipment and agricultural inputs; household foodstuff consumption and clothing; short-term savings and repayment of debts and loans; school fees; socio-cultural activities (weddings, festivals); demerit goods (alcohol and tobacco); and improved family and animal health (Sandhu 1980a, b, c, 1981a, b, c, d).

The economic conditions of large and very large farmers improved even more spectacularly from 1980 onwards. In 1980–1981, this group represented only 0.95 % and 0.2 % respectively of suppliers. By 2009–2010, these percentage figures had increased to 5.36 % and 2.39 %, respectively. Percentage of milk supplied by large farmers to Nestlé also soared dramatically from 6.36 % in 1980–1981, to 17.08 %, in 2009–2010, and from 3.06 to 24.39 % for very large farmers for the corresponding period. In 1980–1981, these two categories supplied 9.42 % of Nestlé's milk purchase. By 2009–2010, this combined figure increased to 41.47 %, nearly a four-fold increase. This is shown in Table 5.3. Viewed from any perspective, this was a very impressive increase in their milk production capabilities, which can be considered to be a proxy for their improving social and economic conditions.

This significant shift in the conditions of large and very large farmers started from about 1985 when they realized the tremendous economic benefits they could obtain in terms of milk production and thus economic returns by switching from buffaloes to cows as dairy animals. This is because the annual milk yield of a female buffalo is about 1,600 kg, with a fat content of about 7 %. However, it is not possible to improve the milk yield of buffaloes by crossbreeding. In contrast, cows could be made significantly higher yielding by crossbreeding them with Holsteins and Jerseys. While the fat content of this milk was low (around 4 %), the milk yield could be increased by 2.5 times and the lactation period could be as high as 300 days per year.

When large farmers became aware of the economic profitability of cow milk production compared to that of buffaloes, they started changing the composition of their herds. Information provided by Nestlé was critical in assisting farmers in

Table 5.3 Changing social and economic conditions of milk suppliers, 1980–1981 to 2009–2010

Farmer's category daily milk production/land holdings	% of total number of suppliers				% of Nestlé's total purchase			
	1980–1981	1991–1992	2003–2004	2009–2010	1980–1981	1991–1992	2003–2004	2009–2010
Farm labourers <2 kg milk; no land	27.31	19.71	17.19	18.5	6.53	3.06	2.17	1.95
Marginal farmers 2–5 kg milk; 0.8 ha	40.5	32.12	28.73	27.47	27.74	14.67	10.56	8.8
Small farmers 5–10 kg milk; 0.8–2 ha	22.56	26.52	26.03	24.94	31.86	25.48	20.22	16.97
Average farmers 10–25 kg milk; 2–4 ha	8.48	18.16	21.73	21.35	24.46	33.6	35.62	30.81
Large farmers 25–50 kg milk; >4 ha	0.95	2.61	4.82	5.36	6.36	12.76	17.35	17.08
Very large farmers >50 kg milk; >4 ha	0.2	0.89	1.5	2.39	3.06	10.46	14.08	24.39

Source Nestlé records

making this shift of dairy animals. It further encouraged this shift by providing veterinary support for artificial insemination as well as making semen available for high yielding Holsteins and Jerseys. As a result, by 1990, cows contributed to 41 % of milk purchased, compared to 99.5 % of buffalo milk procured in 1980. This increasing trend to opt for cow milk production has continued. By 2009, 65 % of milk procured came from cows.

As previously stated, intensified milk production and higher cattle numbers have come about without any land or crop planting trade-offs. In 1992, the size of the herds of regular milk suppliers to the Moga factory had grown by 50 %, with the consequential increase in milk production, without having to make any significant changes in land use practices. Animals have been accommodated in existing plots as land holdings and other resources are more efficiently used and new growing and stock-raising methods are adopted. Not only has the company played a vital role in encouraging farmers to acquire higher milk-yielding animals, it has regularly provided the essential technical support and facilitated financial assistance to make this changeover possible.

Figure 5.1 shows the steadily declining share of buffalo milk in the Moga area during the period of 1980–2012. Clearly, the work of Nestlé has contributed significantly to modernizing the traditional practices of the Punjabi farmers during the past three decades.

These productivity improvements are even more remarkable given India's overall annual milk yield per dairy animal and milk production per capita. In 2001, India reached a production volume of 84 million t of milk, which positioned it as the world leader in the absolute production of cow and buffalo milk. Yet, milk yields per animal are still considerably lower than in other countries with similar

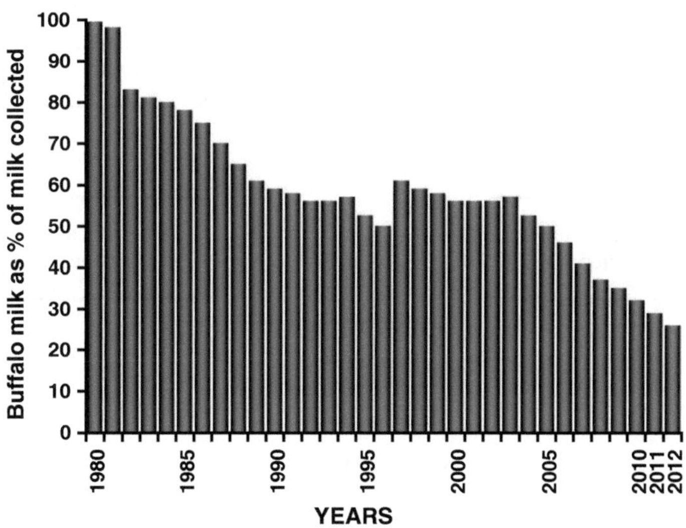

Fig. 5.1 Buffalo milk as % of milk collected. *Source* Nestlé records

production volumes, and so is per capita milk production and consumption. India has three times as many dairy animals as the USA and yet, yields are only about one-tenth of the levels recorded there and around one-fifth of the average annual yields in New Zealand (World Bank 1999; Hemme et al. 2003). These comparative figures point at a significant window of opportunity available to farmers in terms of productivity gains. There is ample room for milk producers to boost yields and income, thus making their own individual farms, and the entire dairy sector, gradually and steadily more competitive and productive, both nationally and internationally.

Nestlé's interventions in Moga have resulted in a series of cumulative benefits, long-term opportunities, and lasting changes in livelihoods that have directly contributed to the economic and social uplifting of the farmers in the region. The company planned and executed a coordinated series of concurrent and long-term interventions that created a virtuous cycle (Fig. 5.2).

Taken all these actions together, it shaped the enabling conditions that were directly responsible for the steady economic progress of a considerable number of marginal farmers to small farmers, and also small farmers to average farmers who had 2–4 ha of land and 3–5 buffaloes or cows. As they progressed further up the economic ladder, their average milk production and thus the milk procurement of Nestlé rose steadily. Consequently their incomes and quality of life improved as well.

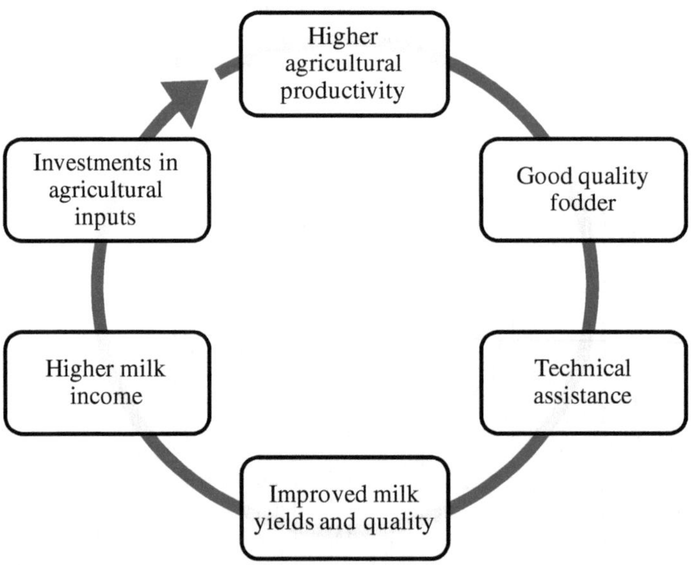

Fig. 5.2 Virtuous cycle created by Nestlé at Moga

5.2 Milk Prices and Milk Economy

From the time the Moga factory opened in 1961 and until India's national economy was liberalized, Government-managed cooperatives regulated prices at which buyers could procure fresh milk. Aiming to offer farmers economically attractive milk prices, Nestlé used the State-regulated rates as floor prices and generally paid a premium of 9–12 % higher. Additional bonuses between 7 and 10 % were introduced to encourage farmers to feed their animals better during the low season, thereby reducing milk supply fluctuations throughout the year. As time progressed, and as farmers appreciated the economic attractiveness of commercial dairying activities, much of north Punjab became a milk-producing region. With expanding milk yields, numerous other milk processing factories were established in the region. By the early 1980s there were at least 17 such processing factories in this area, the majority of which were cooperative enterprises run by the State.

During the early years of the Moga factory, the State constantly intervened in many small but specific and business-shaping decisions, for example, the regular adjustment of milk prices. In 1991, the situation changed dramatically due to new national economic policies that had as a main objective promotion of competition. Following this liberalization, Nestlé set its milk prices considering those paid by the Ludhiana Cooperative (the biggest competitor in the region) as the base rate, and offered its farmers about 5–6 % higher. The average milk price paid by the Moga factory between 1980 and 2012 is shown in Fig. 5.3. Figure 5.4 shows total annual milk procurement by Nestlé from all available sources.

With steadily rising milk prices (Fig. 5.3) and an increasing trend in milk procurement, Nestlé's economic contributions to the milk farmers of the region have been substantial. Just one decade after the factory had opened, the milk

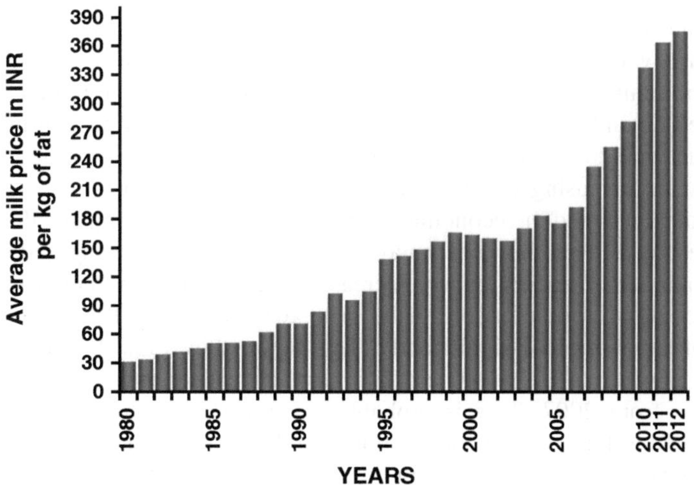

Fig. 5.3 Average milk price in INR per kg of fat. *Source* Nestlé records

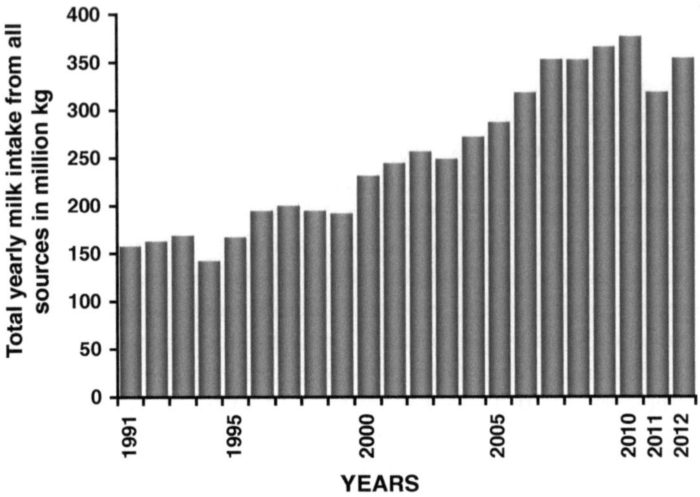

Fig. 5.4 Total yearly milk procurement from all sources in million kg. *Source* Nestlé records

culture was established so soundly in Moga that the total purchase grew over 10 times. Whilst the milk farmers of the region earned INR 910,000 in 1962, this sum had increased to slightly over INR 25.68 million by 1972. In addition, agents in the village diaries were paid over INR 1 million, and milk transportation firms also received receipts for over INR 1 million in 1972. In other words, in few ten years, Nestlé had already injected over INR 27 million into the region's economy only by buying and transporting milk.

This was a completely new type of income-generating activity for the region that simply did not exist a decade earlier. There is no question that the new funds injected into the Moga community contributed to significant individual and collective improvements in the living standards of a large number of households and of the community at large. As overall disposable income in the region increased, so did the demands for other services and goods. This resulting multiplier effect also meant that local entrepreneurs emerged to seize the opportunities being offered by an increasingly thriving and prosperous community.

As farmers realized the economic advantages of dairy farming and the extent to which these additional earnings improved their standard of living and quality of life, the number of milk suppliers increased steadily. This is shown in Fig. 5.5, which provides information up to the year 2012. Between 1995 and 2000, the number of suppliers rose significantly. This figure shows that between 2000 and 2006, the number of suppliers seems to have plateaued, and the trend has been downwards from 2007. This is probably because during the post-2000 period, Nestlé started to face intense competition from both public and private sector milk procurers who started buying milk in Punjab, Rajasthan and Haryana for their own commercial requirements. Many of these new competitors are offering farmers

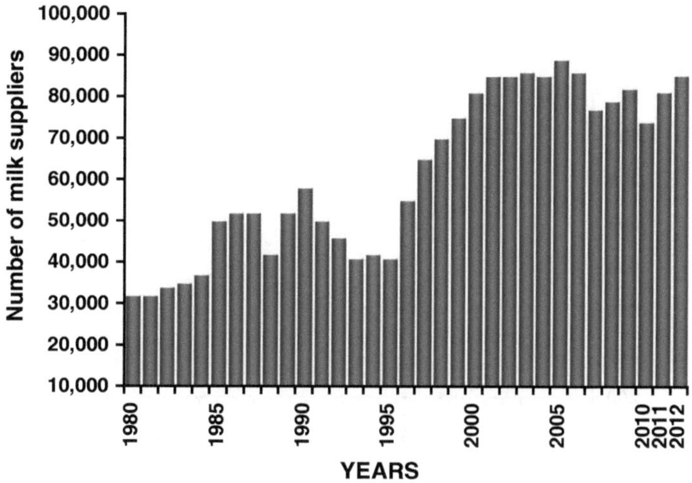

Fig. 5.5 Number of milk suppliers. *Source* Nestlé records

higher wholesale milk prices compared to Nestlé but they do not provide as many of the additional services Nestlé does. Accordingly, direct price comparison alone would give an incorrect view of the current situation.

Collection centres and dairying processing facilities are important infrastructural elements in the sector and are also closely intertwined to the number of supplying farmers. Naturally, as the number of farmers supplying milk to Nestlé has grown steadily for most of the time since the Moga factory opened, the number of milk collection centres has increased as well. The number mostly increased till 2008. Thereafter, it has been somewhat inconsistent, reflecting intensified competition for milk in the region. This is shown in Fig. 5.6.

Nestlé has absorbed and continues to absorb, many of the internal and external costs for milk production of the farmers, thereby extending many direct and indirect benefits to them. The company has taken on a broad range of roles to meet the demands, needs and requests of its suppliers, workers, supporting businesses, producers and costumers alike. For instance, milk rates paid by Nestlé do not reflect the reputational gains farmers derive from being associated with the company. Those producing milk for the company are known for their expertise in dairying activities. They must follow stringent quality and hygiene standards, and fulfil strict food safety, quality and processing requirements which are clearly enunciated by the company and are strictly enforced. Equally, their reliability, business ethics, professionalism and integrity must be beyond reproach.

When competing dairy firms arrive in the region aiming at capturing a share of the local milk producing market, they are counting on finding and drawing in from the pool of skilled and experienced farmers and supporting industries in Moga. Moreover, the many ancillary firms that have thrived in the region constitute the necessary supporting infrastructure to provide the services and inputs that help

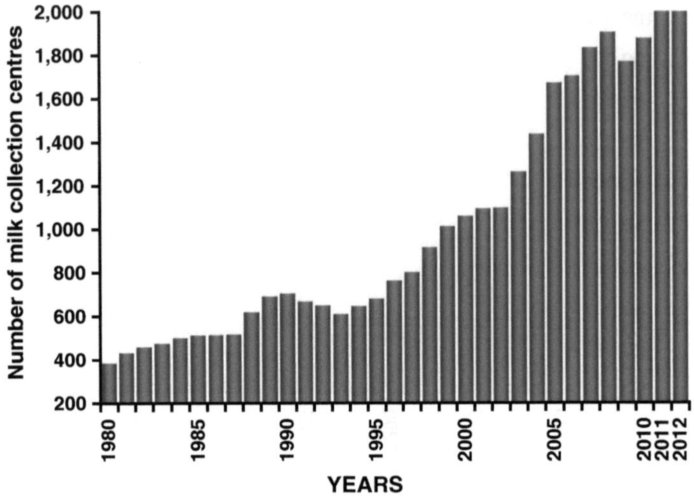

Fig. 5.6 Number of milk collection centres. *Source* Nestlé records

form a successful milk supply chain. Many of these prerequisites have been the result of the dairy sector infrastructure Nestlé has established and the human capital investments the company has made over the decades through its Agricultural Extension Services.

References

Delgado C, Narrod CA, Tiongco MM (2003) Policy, technical, and environmental determinants and implications of the scaling-up of livestock production in four fast-growing developing countries: a synthesis. Phase I. An IFPRI-FAO Project. Available via DIALOG. http://www.fao.org/WAIRDOCS/LEAD/X6170E/x6170e00.htm#Contents

Hemme T, Garcia O, Saha A (2003) A review of milk production in India with particular emphasis on small scale producers. PPLPI Working Paper No. 2. Food and Agricultural Organization, Rome

Nestlé India Limited (1991) 32th Annual Report, 1990, New Delhi, Directors' and Auditors' Report

Sharma VP, Vir Singh R, Staal S, Delgado C (2002) Critical issues for poor in the India dairy sector on the threshold of a new era. Livestock Industrialisation Project (Phase). An IFPRI-FAO Project, Food and Agriculture Organization, Rome

Sandhu, TS (1980a). Impact of Moga Milk Plant on Agricultural & Rural Development. Food Specialities Limited, Moga

Sandhu, TS (1980b). Anatomy of Milk Sales. Food Specialities Limited, Moga

Sandhu, TS (1980c). Human Factor & Milk Forecasting. Food Specialities Limited, Moga

Sandhu TS (1981a) Extent of full employment generated by Moga Milk Plants in its milk shed area. Food Specialities Limited, Moga

Sandhu TS (1981b) Evaluation of development activities of Moga Milk Plant. Food Specialities Limited, Moga

Sandhu TS (1981c) Inputs for milk procurement. Food Specialities Limited, Moga
Sandhu TS (1981d). Prosperity? Food Specialities Limited, Moga
World Bank (1999) India Livestock sector review: enhancing growth and development. Allied Publishers, Mumbai

Chapter 6
The Key to Success: Agricultural Extension Services

Driven by the need to increase the quantity and quality of milk, and motivated by the importance that attaining greater social objectives has in securing the company's supply base and in increasing productivity in the value chain, Nestlé and its supplying farmers have established a 'community of interests'. The company encourages, supports and facilitates those factors necessary for its own successful operation. Nestlé's Agricultural Extension Services (AES) have been a defining factor in securing adequate and regular inputs and raw materials. As such, the company has made sizeable investments in staff and farmers' training and capacity building as part of its efforts to attain factory success and further rural development. These efforts to address skill and knowledge gaps among milk producers were extensively addressed from the very onset of the factory's operations. As early as in 1968, the company had recorded over 150 audio-visual materials and educational films that were shown to over 50,000 farmers that year alone (Food Specialities Limited 1968).

These services operate as a continuous and expanding network of free, relevant and cutting-edge veterinary, animal husbandry, dairying and agricultural advice, technical assistance, education and capacity building sessions. Using endogenous knowledge and partnering with local institutions and universities, Nestlé has supported, fostered and disseminated innovations, and has encouraged their application. The number of areas where it has promoted novel methods and technologies has steadily augmented with the expansion of the range of goods it produces in the Moga factory. This robust and dependable support system has also made farmers less risk-adverse when it comes to adopting new technologies, implementing new techniques, planting new crops or putting into practice recently acquired knowledge. Unsurprisingly, continuous education and training have created confidence and credibility in Nestlé, especially since all these investments in human capital have remained mindful of local conditions and are context-sensitive.

These efforts to capitalize on existing available knowledge and foster and disseminate innovations have been consistently supported through partnerships with local institutions. For example, numerous schemes have been designed and implemented in conjunction with the Punjab Agricultural University (PAU). This

A. K. Biswas et al., *Creating Shared Value*, SpringerBriefs on Case Studies of Sustainable Development, DOI: 10.1007/978-3-319-01463-0_6, © The Author(s) 2014

relationship has resulted in lectures being delivered by University staff, training sessions for farmers to take up responsibilities in the milk district; face-to-face discussions to disseminate new farming and breeding techniques, and to give maintenance to farm machinery; as well as the production of newsletters and other publications on animal husbandry and agricultural issues. In 1992 alone, 29 local students graduated from PUA and other agricultural universities and joined the Extension Services team. Moreover, the Punjab Agricultural Husbandry Department has organized a series of specialized welfare camps to solve complicated veterinary cases and one-day trips to nearby agricultural colleges for farmers to become acquainted with the latest farming techniques and how these have been adapted to local conditions (Nestlé 1993; Nestlé India Limited 1990, 2011; Sandhu 1984, 1985a, b).

Amongst the many benefits the assistance services Nestlé delivers to milk producers and farmers, some of the most noteworthy have been attained in the area of animal health. As mentioned above, this support has been prepared and delivered in partnership with the local universities. It also has reached to a growing number of farmers. When the company opened the Moga factory, calf mortality rates were as high as 60 %. Ten years later, these had fallen to 25 %. They further dropped again to 15 % by 1992 (Nestlé India Limited 1990; Nestlé 1993). In 1962, around 50,000 vaccinations were applied and by 1989 this number had gone up to 132,700 shots, mostly against food-and-mouth diseases. That same year, more than 100,000 dairy animals received routine veterinary attention. As a result of these sizeable and joint efforts, worm infestation diseases have been eradicated, fluke has disappeared, and haemorrhagic septicaemia has been halved (Nestlé India Limited 1990; Sandhu 1978, undated). This has significantly improved livestock health and thus has enhanced milk productivity.

More importantly, extension services offered by Nestlé have been decisive in sustaining and further elevating production rates and thus income from milk sales, especially during drought years. Normally, a failed or low harvest would bring about much economic hardship, indebtedness and even compromise food security of individual families and sometimes even of an entire village. Instead, technical assistance has helped milk producers keep procurement levels at constant and ever increasing levels, thereby guaranteeing a steady and reliable source of income. This has helped to cushion fluctuations in the market price and harvested volume of the crops the region produces (Food Specialities Limited 1968, 1970).

Nestlé managers have also been mindful of the best didactic and educational strategies that are best suited to cater to the needs of the Moga's rural population. Accordingly, group activities have been favoured and extension programmes organized through field visits, village camps, seminars, models and demonstrations, literature in local language, etc. These schemes have covered issues related to animal husbandry, pesticide residues reduction, improved crop cultivation practices, clean milk production, efficient utilization of fuel, biogas production and other appropriate issues. Within the first two decades of the Moga factory, about 20 such programmes were organized each year, with an average participation of 300–400 farmers. During the field days, groups of farmers meet with Nestlé experts to discuss profitability, animal health and breeding as well as milk quality issues.

Moreover, audio-visual training has helped to improve farmers' knowledge base. The literacy rates are still low. The training programmes now rely on pictorial illustrations, practical lessons, radio shows and pre-recorded tapes. Outreach efforts have also relied on the adoption of innovative dissemination and message-reinforcing methods. For example, envelopes holding the farmers' payments have been printed with key messages from the technical training programmes. This same method has been used to make public recognition of successful farmers, communicate achievements to the larger body of milk suppliers and encourage others to adopt similar good practices.

As the knowledge and economic conditions of the farmers have increased, they have been able to improve the number and/or quality of their herd assisted by the support services received from Nestlé, either freely or at cost (for example, veterinary medicines or cattle-feed). Agronomic advice was given on how to use existing land holdings ever more efficiently. Because of the new growing and stock-raising methods, cattle numbers have grown without additional land acquisitions or crop planting trade-offs as animals could be accommodated in and fed from existing plots. Buffaloes and cows literally became 'cash cows' which ensured a regular, steady and predictable cash flow that contributed to about 40 % of their annual income. An average farmer used about 25 % of available land for fodder which meant that 75 % of the plot was available for growing traditional crops which could be used by the households or sold to generate cash incomes.

The company also offers systematic and/or specialized factory training, apprenticeships and postings in other factories abroad (see Annex 4, Case Study 1 describing Mr Jatinder Singh's experience with Nestlé). Members of Moga's Agricultural Extensions Services receive continuous technical assistance at the Agricultural Services Department in Nestlé Headquarters in Vevey, where information and knowledge from the company's worldwide markets is centralized. At the same time, agricultural experts from the Vevey office periodically visit the Moga factory as part of a concerted cross-fertilization and knowledge and know-how exchange scheme (Nestlé 1993). Additionally, the company carries out regular research projects and investigations to track progress; identify areas of improvement and commonly encountered challenges and difficulties; and monitor socio-economic changes in the community.

With these initiatives, the company supports research, innovation, development, dissemination and adoption of successful practices regarding key inputs, raw materials and innovative farming methods. Moreover, Nestlé has promoted peer-to-peer learning among suppliers and publicly recognizes successful farmers, publicizing their achievements and encouraging other entrepreneurs to follow suit. Such efforts are also extended to ancillary farms as specific programmes have been designed and implemented to help close farmers' skill gaps affecting the supply of raw materials to Nestlé. Company staff is oftentimes extensively involved at its suppliers' locations to ensure the successful implementation of new technologies and quality control processes. This is discussed in detail later in the book for two main ancillary partners Nestlé has at Moga: Paras Spices Private Limited and the milk transportation company Brar Enterprises Limited.

Since operations in the Moga factory began, plant directors were prompt to notice that the bulk of the work with the animals is done by women, and thus their continuing education and training are needed to ensure that health, nutrition and cleanliness requirements are consistently met in the families of Nestlé's suppliers. Long-term rural development is not possible, or sustainable, without the active involvement and the participation of women.

6.1 Nestlé and Women

During the early 1960s, when agricultural practices were more labour-intensive and barely enough to cover subsistence level needs, women played vital roles in maintaining home and farm economies. Even now, from dawn to dusk, village women work continuously in farm-related activities, carrying out household chores, picking crops, managing livestock, taking care of family members, including children and elderly, and other non-farming work such as spinning, weaving, sewing, etc. Nevertheless, the vast majority does not receive any monetary remuneration for their work, and more often than not, are not given full credit for the activities they carry out and the important productive role they play in household, rural, community and regional economies.

Similar to other places in India, farming and dairying activities in and around the Moga area are sharply divided along gender lines. An early Nestlé study surveyed 1,298 farming families from seven villages to obtain information on how buffalo husbandry tasks are distributed. The survey indicated that 87 % of the milking was done by women, as well as taking care of cutting and chaffing fodder; feeding, watering, cleaning and care of the animals; and animal shed cleaning. A smaller percentage of women was involved in delivering milk to collection centres. The livestock husbandry labour divisions in Moga reflect the reality seen elsewhere in the country. It has been estimated that 75 million women are engaged in dairying as opposed to only 15 million men, thereby constituting around 71 % of the workforce in livestock farming and animal husbandry (Singa 2007). It is thus evident that productivity, quality of milk and hygienic practices can only improve further significantly if women receive additional appropriate training. If productivity in the milk value chain is to be improved, social attitudes to gender and women farm labour have to be reshaped in the region.

Women's conditions have changed positively over the last few decades especially after the arrival of Nestlé at Moga. Yet, a lot remains to be done. Education and literacy rates are particularly important factors if women are to be empowered. They also affect and shape how the company can communicate with farmwomen and the type of programmes that should be specifically devised to address their perceived and future specific needs. Literacy rates in Punjab have been on the rise over the last decade and are considerably higher than the national average of

74.04 % recorded in the 2011 census. In 2001, 69.65 % of Punjab's population was considered to be literate. However, the female literacy rate stood only at 60.53 % compared to 79.66 % among males. Ten years later, 76.68 % of Punjabis were considered to be literate, with male literacy standing at 81.48 % whilst female literacy reached 71.34 %. Nevertheless, urban and rural divides are considerable. By 2011, in rural areas of Punjab, the average literacy rate was 72.45 %, but 77.92 % for males and significantly lower, 66.47 %, for females (Indian Census 2001, 2011).

In spite of improving literacy rates witnessed in recent years, women's access to formal education has continued to remain low. Accordingly, imparting knowledge still has to rely heavily on audio-visual media, like diagrams, charts, illustrations, posters, slides, photos, pictorial guides and short films. These audio-visual materials that were specially prepared by Nestlé to educate and train women milk farmers about improved dairy practices have reached well over 56,000 female farmers in and around Moga area by 2012.

Participation in remunerative activities has had a particularly relevant and positive impact on women. As their role in dairying activities has become increasingly recognized, so has their direct and indirect involvement in the industry. The Village Women Dairy Development Programme focuses on cattle health, good feeding and breeding practices, improved water use in the farm and agricultural practices. The Healthy Kids Programme focuses on nutrition. Additionally, farmwomen, in batches of 60–70, were brought to the Moga Factory where agronomists and veterinarians discussed appropriate issues related to animal husbandry.

Higher household income does not necessarily translate in improved nutritional and health for women and children when those earnings are controlled by men. As women in Moga have joined the formal and remunerated labour force, their relative bargaining power in the household has meant that those income receipts are more likely to be used to guarantee good child health and nutritional security, augment investments on education of the children, specially girls, and help raise welfare and living standards. Some community members have even reshaped gender expectations by taking on professional roles in traditionally male-dominated activities such as tractor driving.

It can be confidently concluded that nutrition and education levels have improved in the households of those 56,000 women who have received formal training and education from Nestlé's staff at the Moga factory. In turn, this has had positive impacts on the health of the Nestlé suppliers as well as on their animals.

Overall, new and enhanced agricultural, veterinarian, managerial and trade-related knowledge has gone well beyond closing farmers' skill gaps, changing attitudes to milk trading, engaging women in dairying activities and opening up opportunities for non-agricultural rural employment. All these developments have positively impacted on the quantity and quality of raw materials and services Nestlé needs to operate a successful dairy products industry. With a deeper and larger pool of capacities and expertise at their disposal, farmers have enlarged their occupational flexibility and mobility. Oftentimes it has accelerated and/or intensified their engagement in other economic activities that are not necessarily tied to agriculture. This is because the transferable and practice-based skills they have acquired allow them to perform new and different tasks as they evolve and put their newly acquired capacities to use.

Such capacity building schemes highlight relational over transactional interactions. Experts, agronomists and agrotechnicians delivering the extension services spend most of their time, around 90 %, in the fields with the farmers. They have regularly kept direct and close contact and rapport with over 85,000 individual farmers. These schemes rely on the establishment and strengthening of mutual and effective communication avenues between the company and supplying farmers, farmers and veterinarians, farmers and agronomists and farmers and extension departments (Food Specialities Limited 1980; Nestlé 1993).

This study found the importance and key role of the company's Agricultural Extension Services. From the very onset of the factory's operations, farmers supplying milk to Nestlé have received free expert veterinary advice, medicines for animals supplied on a bulk-cost and non-profit basis, special animal feed exclusively manufactured by a company selected by Nestlé according to its own specifications and where the production is closely and carefully supervised by the company to ensure no adulteration takes place; and more crucially formal training and capacity building to intensify production and maintain quality. Such initiatives have traditionally responded to local needs and are tailored to address issues of relevance to the Moga community. They have also played a crucial role in maintaining and boosting production even in years when, due to climatic conditions, harvests have been poor and so have incomes from agricultural activities. In the last 50 years, milk payments have reduced farmers' vulnerabilities as they compensate for crop losses during drought years. The company has invested and reinvested in furthering the skills of Moga's milk producers and those collaborating in the entire infrastructure supporting the dairy sector.

6.2 Milk Collection and Role of Agents

Agents in milk collection centres have been one of the cornerstones of Nestlé's success. With more than 50,000 milk suppliers, the tasks of the milk collection agents to manage the process regularly, efficiently, transparently and reliably are vitally important. They have played a key role in helping manage the area's highly fragmented milk supply chain.

Agents collect milk twice a day, check its quality, issue receipts to suppliers and keep the collection open until milk is collected by the vans and delivered to the Moga factory. The agents then return to the field, shops or other activities. Although paid on a commission basis, Nestlé's agents are not simply milk dealers. In fact, they constitute a vital link between milk supplying farmers and the company's agriculture and veterinary extension services run by its Agricultural Extension Services.

Since 1961, when the Moga factory was built, its agents have maintained twice-a-day communication avenue with the villagers in addition to being a role model, and a channel to obtain information on necessary and innovative programmes. They have had the responsibility to share with the milk producers and farmers the importance of implementation of Nestlé's expert advice related to agriculture, animal health, dairy practices and quality standards. Before telephones, mobiles and computers became widely available, the agents served as the essential communication links between the company and its milk suppliers on a regular basis in otherwise what was then scattered and somewhat remote communities.

Agents were vital for the accurate measurement of milk in a place and society where people were used to the business practices of mostly unreliable and unscrupulous *dodis*. They calculated fat contents and ensured timely, regular and correct payments were made to the supplying farmers. During the past 50 years, by working fairly and transparently, these agents have in effect become an extension of the company's personnel. Their own villages have also appreciated their contributions and awarded them a high social profile. Agents are often publicly recognized as honest, reliable, professional and well-connected community members. They have become role models, with their success stories being repeated once and again in public events and village gatherings.

The roles of the agents are fundamental to the business model Nestlé has put in place in Moga and elsewhere. They are selected based on nine attributes: work, honesty, attitude, dedication, character, authority, political neutrality, ambition, and ability to create and nurture public relations (Sandhu, undated). Because of their importance, they have to be chosen carefully and then cultured and trained to serve as an efficient and effective link between the milk suppliers and the company. Agents have been of extraordinary help to both the sides, communicating each other's views and requirements, disseminating information, voicing concerns, dissipating doubts, transmitting successful practices and bridging the gap between the two groups in all areas.

Since agents were first hired, their performance has been reviewed regularly. The selection and evaluation process has been transparent and has remained serious enough to instil confidence in the milk suppliers. Consistent with an overarching responsible and competitive corporate model, Nestlé has consistently tried to keep the best performing agents. Equally, it has also replaced those who did not perform well. From 1962 to 1980, 341 agents were replaced and 113 were reshuffled. This is a very large number considering that until 1970, the number of agents was around 130, increasing to 300 during 1970–1980 period. However, this regular review of the agents' performance has had salutary impacts on their overall effectiveness since they are aware that they are being closely and regularly monitored.

Nestlé managers realized very early that personal interactions with the farmers were essential to develop a successful milk production area. The company's experience in various villages such as Dhudike, Chuharchak, Bhinderkalan and Daudhar, for example, suggests strong investment in human resources development, including agents, making checks and balances a regular practice. For instance, initially, agents and farmers used to adulterate milk by adding water, as the commission they received was not linked to quality. However, very soon, Nestlé found out about this unacceptable practice and decided to change the rules, linking the agents' commission to the quality of the milk received. Management practices have been tailored since that time responding to the different situations, with important steps being taken to motivate both agents and farmers. Nestlé's public relations programme started in June 1971 and has steadily contributed to improving communication and outreach practices. Also, massive extension education schemes were launched around that time.

Today, Nestlé's agents are well respected in the villages and are regarded as efficient and trustworthy persons by both the farmers and the company. Agents enjoy not only status but also a larger social network that makes their bonding with their 'clients' stronger, and thus more effective. The interactions they hold with the supplying farmers go well beyond simple monetary transactions and merchandise exchanges and are based on long-term relational and social capital partnerships. They have developed personal relations with the farmers and offer them technical advice and inputs as well as information on animal health care. Some agents have even helped farmers financially with their own resources. As noted earlier, the personal interactions and rapport have been, and continue to be, vital factors to develop and maintain the milk production area in Moga, and thus to Nestlé's success (Photo 1).

Photo 1 Interviewed milk suppliers to agency No. 1, Dharamkot village

References

Census-2011 (2011). Punjab population census data 2011, India. Available via DIALOG. http://www.census2011.co.in/census/state/punjab.html
Food Specialities Limited (1968) Report and Accounts for the Year ended 31st December, 1967, Directors' and Auditors' Report, New Delhi
Food Specialities Limited (1970) Report and Accounts for the Year ended 31st December, 1969, Directors' and Auditors' Report, New Delhi
Food Specialities Limited (1980) Annual Report, 1979, Directors' and Auditors' Report, New Delhi
Nestlé (1993) Nestlé in India 1962–1992, Nestlé, Vevey
Nestlé India Limited (1990) 31st Annual Report, 1989, Directors' and Auditors' Report, New Delhi
Nestlé India Limited (2011) Annual Report, 2010, Directors' and Auditors' Report, Gurgaon
Sandhu TS (1978) Agro-climatic and Other Situations. Yearbook. Food Specialities Limited, Moga
Sandhu TS (1984) Constraints in milk production and their removal. Food Specialities Limited, Moga
Sandhu TS (1985a) Challenges for milk procurement and extension workers. Food Specialities Limited, Moga
Sandhu TS (1985b) Milk development of milk shed area. Food Specialities Limited, Moga
Singa OP (2007) Agro-industries characterization and appraisal: dairy in India. Dairy Farmers' Organization, Anand, India. Working paper No. 21. Food and Agricultural Organization, Rome

Chapter 7
Contributions to the Local Economy

Nestlé's work to make of Moga a thriving milk district is perhaps one of the best examples of how the goals, success, needs and demands of different community of interests are closely interlinked. Without claiming to be a development agency, Nestlé nevertheless consciously made crucial contributions to the community of Moga on issues that were fundamental for the socio-economic development of the region as a whole, and also to its farmers at the household and individual levels.

7.1 Tax Payments

Table 7.1 shows the various tax contributions the Moga factory has paid to various levels of Government, including taxes for central sales, value added, entry of goods into local areas of use or consumption, the purchase of milk, cess on liquid milk, etc. Within a span of 15 years, between 1996 and 2011, total taxes paid increased over 13-fold from INR 42.60 million to INR 556.52 million. A significant percentage of this amount is administered by local and Punjab authorities, thus directly contributing to the development of the area and the delivery of services in the Moga district and the villages around it. The total annual taxes paid by Nestlé to the Central, Punjab and municipal levels of Government for the period of 1996–2012 are shown in Fig. 7.1.

An important way to assess the impacts of the Nestlé factory in the development of the Moga municipal district is by considering the Octroi taxes imposed directly by the local authorities. These duties were "levied on the entry of goods into local areas of consumption, use or sale therein" (Deep 2002: 265) and constitute an important, steady and reliable source of income for Indian local bodies and municipalities.[1] Between 1997 and 2006 (in 2007 this tax was drastically

[1] The Octroi rate varied from item to item according to a list prepared and published by each state. Up to 2002, this tax constituted around one-fourth of the total tax revenue of local bodies in India and it was particularly important for the tax structure in Haryana, Orissa, Punjab, Rajasthan, Gujarat and Maharashtra. For a detailed account on the merits of the Octroi tax in Punjab see The Octroi Debate: The Case of Punjaj by Gagan Deep, in Singh 2002.

A. K. Biswas et al., *Creating Shared Value*, SpringerBriefs on Case Studies of Sustainable Development, DOI: 10.1007/978-3-319-01463-0_7,

Table 7.1 Contribution to Government taxes by Nestlé

Description	1996	1997	1998	1999	2000	2001	2002	2003	2004	2005	2006	2007	2008	2009	2010	2011	2012
Central sales tax	4.48	2.79	3.15	3.31	4.75	7.00	6.00	3.95	2.46	2.4	3.1	5.09	9.19	3.35	3.75	3.35	4.3
Sales tax	15.40	17.68	18.84	22.84	29.29	39.40	39.60	46.39	47.74	15.8	–	–	–	–	–	–	–
VAT	–	–	–	–	–	–	–	–	–	55.4	85.96	122.17	162.06	187.35	196.26	197.07	237.7
Octroi	Not available	20.00	18.00	21.30	20.50	20.00	15.70	25.00	29.42	33.5	23.98	0.37	0.54	1.26	0.82	0.48	–
Entry tax	–	–	–	–	–	–	–	–	–	–	–	–	–	28.55	85.13	107.49	–
Works contract tax	–	–	0.41	0.25	0.59	0.40	0.40	0.40	0.40	0.20	0.30	0.43	0.64	0.99	0.59	2.90	6.1
Purchase tax on milk	19.93	30.83	74.79	75.00	18.75	–	4.79	–	144.90	83.1	98	124.36	133.51	135.34	225.32	245.23	296.6
Purchase tax others	2.71	2.40	1.80	1.16	0.82	1.10	1.40	2.31	2.45	0.93	–	–	–	–	–	–	–
Cess on liquid milk	–	–	–	–	12.45	27.38	27.38	–	–	–	–	–	–	–	–	–	–
Total	42.60	73.71	117.00	123.87	87.15	95.28	95.26	78.05	227.37	191.33	211.34	252.42	305.94	356.84	511.87	556.52	544.7

Source Nestlé records

– During the period 19th July 2000 to 11th September 2002, cess on milk was imposed and purchase tax was abolished

– The information on the Value Added Tax, VAT, on Nestlé products manufactured outside Punjab but sold in Punjab is awaited

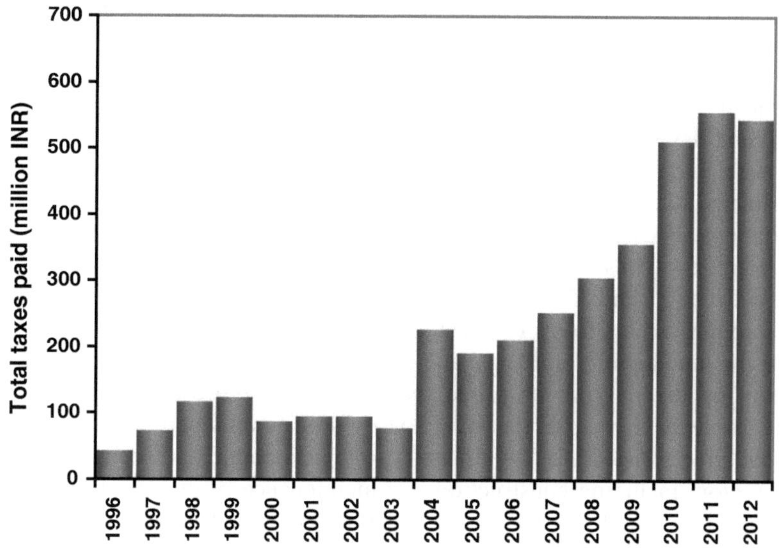

Fig. 7.1 Total taxes paid by Moga factory to central, state and municipal authorities, 1996–2012. *Source* Nestlé records

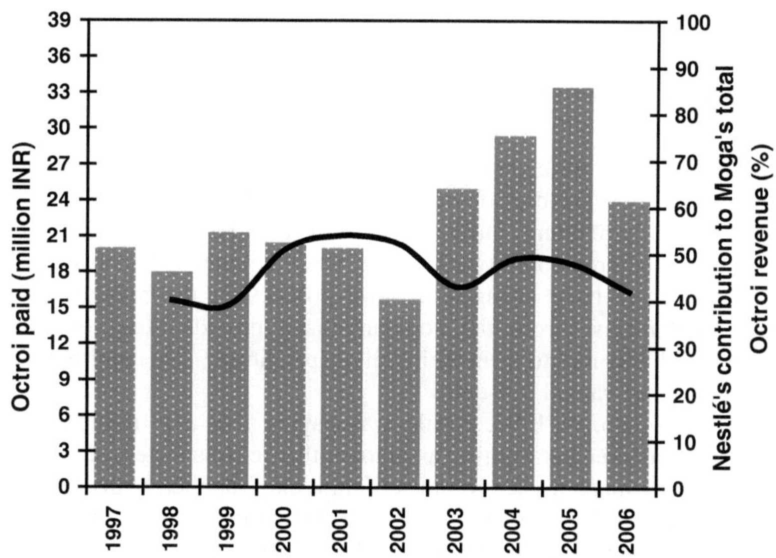

Fig. 7.2 Octroi paid to Moga municipality, 1997–2006. *Source* Nestlé records

changed), Nestlé's tax contribution represented 20–35 % of the municipality's total income. This is shown in Fig. 7.2. The company was by far Moga's most important single tax contributor. Thus this economic support undoubtedly underpinned much of the Moga's public services and Government-initiated and funded development activities.

With the radical changes in the structure and scope of Octroi taxes from 2007 onwards, it is not possible to estimate directly the contributions of Nestlé for the upkeep and development of the Moga municipality. This is primarily because the taxes were paid to the Central and state level Governments instead that directly to the municipality. Moga, in turn, received support from the State and the Central Governments under different forms. We estimate that the Nestlé factory and its direct suppliers (considering only Nestlé-related activities) have contributed anywhere between 25 and 35 % of the total income of the Moga municipality in recent years through various forms of taxes. These fiscal payments have added to the tax base of the municipality of Moga, and thus strengthened the resilience of the local economy by diversifying its productive activities. Nestlé's tax contributions in various forms have assured the municipality's financial viability for decades, and no doubt have assisted very significantly to the economic and social development of the region, including construction and maintenance of its infrastructure.

It should be noted that the above discussion does not include the indirect tax payments, or similar taxes, paid by the company's suppliers on Nestlé-related economic activities, which are also likely to be significant. If the Moga factory did not exist, all these taxes, and thus the incomes of the Government at various levels would have been much less, especially for the Moga municipality.

7.2 Farmers' Livelihoods

Absolute milk pay or milk remuneration to producing farmers have provided a stable day-to-day household income for the farmers who have supplied milk to the Moga factory. In agriculture, farmers usually receive payments from their commercially produced crops only twice a year, when produce is harvested and sold. Once these farmers have engaged in milk production, the regular fortnightly payments they receive from the company have provided the necessary and smoothened cash incomes to meet their daily expenditures, also contributing to the households' overall capital resources.

Twenty years ago, in 1992, Nestlé was paying INR 609 million a year in relation to the procurement of milk, an amount that reflects the expanding number of farmers and the much higher volumes of milk they supplied to the factory. This amount, as noted earlier, was only INR 910,000 in 1962, when milk productivity was low and suppliers few. It had increased to slightly over INR 25.68 million in 1972 as farmers favoured the adoption of higher-yielding dairying practices and the milk culture was growing in Moga. Also in 1972, the commission paid to the milk agents exceeded INR 1 million, whereas factory employees received payments for about INR 4 million. Nearly all of this money was injected into individual households and the local economy at large and for the most part was spent in the villages where the farmers lived. As overall disposable income in the region

Table 7.2 Milk development expenses for 1979 and 1980 (in INR)

	Amount (1979)	Amount (1980)
Value of milk purchased	89,745,982	110,288,842
Total commissions paid to agents	3,322,382	4,469,885
Total room rents paid to room owners	84,235	141,640
Total transportation expenses on milk	2,515,012	3,599,205
Total transportation expenses for cattle feed mixtures, milk collection expenses, equipment, stores, sundries, etc.	613,570	700,000
Milk inspection vehicles expenses	293,579	291,147
Total staff salary	351,360	435,241
Total amount	97,219,699	119,925,960

Source Nestlé records

increased, so did the demand for other services and goods, triggering a multiplier effect in which a progressively more prosperous community led to the opening of more opportunities for socio-economic development.

Table 7.2 shows the money injected by Nestlé in the local area for milk-related expenses for 1979 and 1980, when milk collection was almost stable. By the end of this period, Nestlé's direct contribution to Moga's economy, for milk procurement alone, was nearly INR 120 million.

Even though these figures only refer to 1979 and 1980, over the years, the amounts Nestlé has disbursed in the Moga region as payments to farmers for milk procurement activities have gone up steadily. In 1992 alone, more than INR 80 million were paid in local taxes, commissions to dairy milk agents, and transportation costs to the drivers covering the different milk routes. All together, in that year, over INR 609 million were injected in the local economy just by purchasing milk. This corresponded to an equivalent annual income for 46,230 workers when paid the minimum legal salary fixed by the Punjab Government. In 2010 and 2011, the company's payments to its milk producers alone exceeded INR 6,120 million each year. This substantial sum has significantly contributed to the amelioration of the standard of living of the people in the factory's neighbouring villages and has also furthered the individual and collective economic and social progress of those living there.

Given the scope of the present study, it was not possible to estimate the catalytic and cumulative effects the injection of these funds has had in enhancing local economic activities. As farmers' disposable income increases so does their spending capacity and the aggregate demand for additional goods and services, which in turn brings in additional resources and economic activities to the local area. Our best estimate is that for every INR 1.00 that Nestlé has injected to Moga's economy, it may have had a multiplier effect of around INR 3.00–3.50 in additional private economic activities.

Viewed from any direction, the company's presence in Moga and the contribution it has directly and indirectly made through wages, commissions, and payments to improve the standard of living and quality of life of the people in this region have indeed been very significant. These benefits, however, need to also take into consideration the additional multiplier effect that has come about from taxes and fiscal dues.

7.3 Employment Generation

It is important to note that the establishment of Moga factory has positively impacted not only milk producers, but also other demographic and occupational groups of people in the area. The socio-economic and cultural changes the presence of Nestlé has triggered in the region have unleashed new commercial and development opportunities. Whilst the milk farmers have directly and significantly benefitted because of the actions undertaken by Nestlé, it should be stressed that the factory has also made significant contributions in generating new and remunerative employment opportunities in the Moga region, and sometimes outside Punjab and even India, by direct and indirect means. In fact, as part of this cluster development seen in Moga, more jobs have been created in the ancillary industries, mostly in the production and transportation of raw materials, than on milk-related activities carried out in the factory itself.

The plant provided direct employment to 2,400 people in 2011. Most of these workers come from Moga itself, which suggests the extent to which the company draws and builds local talent and capacities, and recruits workers from the communities around the factory. This study has further estimated that another 86,371 persons are in full-time employment in support and ancillary industries to satisfy the direct requirements only of the Moga factory.

7.4 Ancillary Firms

7.4.1 Raw Material Suppliers

Nestlé's diversification into the manufacture of non-dairy products has led it to search for regular supplies of good quality ingredients such as peppers, onions, garlic, spices and other vegetables. The Moga factory also has a great demand for printing and packaging materials. Except for few ingredients such as lactose, vitamins and enzymes, Nestlé exclusively focused on getting raw materials locally instead of depending on imported supplies. As early as 1974, Nestlé was already purchasing local raw materials worth more than INR 58.3 million. By the early 1980s, imports were almost non-existent. For example, in 1982, imports constituted only 0.03 % of total value of purchase.

Whilst Nestlé's work with dairy farmers of Moga is generally well appreciated within the region, its work with other suppliers of raw materials such as coffee, spices, chicory and sugar is almost unknown outside the company. For example, the company has consistently helped the local suppliers to regularly increase output and also to improve the quality of their products. It has also encouraged them to grow crops that the Nestlé factories need as raw materials for manufacturing, as has been the case with chicory cultivation.

During the past 10 years, the number of chicory farmers benefiting from Nestlé's technical and capacity building assistance has increased from 1,000 to 7,500. Total production has expanded by 600 % in only one decade, from 2,000 t in the year 2000 to 12,000 t in 2010. Quality levels have remained consistently high. Similarly to chicory producers, many other entrepreneurs have seized the opportunities the opening of the Nestlé factory has made available to supply needed, services and products.

Many ancillary enterprises have developed a symbiotic relationship with the Moga plant, tying most of their success and business expansion plans to that of its business partner. These initial relationships were primarily linked to the Moga factory, but these have now successfully transcended to other Nestlé factories in India, and in a few cases even to the company's manufacturing facilities outside the country and to other firms altogether.

7.4.2 Local Tastes

The most remarkable example of such an industry is Paras Spices Private Limited, a company that in 30 years transformed itself from a small mom-and-pop grocery shop to a company with an annual turnover of more than INR 1,000 million. Nestlé pioneered the 'Indianization' of food recipes such as Maggie Masala to respond to local tastes and needs. This opened up a huge window of opportunity for spice producers, processors and suppliers. Paras built a systematic collaboration with Nestlé, establishing a wide and intensive range of interactions and mutually beneficial investments and training requirements. The company now sells over 70 % of its products to Nestlé, including exporting them to Nestlé factories in South Africa.

For the first two decades of operation, the Moga factory procured the spices it required from south India, but managers were confronted with the serious problem of inconsistent quality. It was in 1982 when Nestlé had to buy 2 kg of turmeric on an urgent basis. They purchased it from Paras, which was then a small family-owned grocery shop in Moga. The quality of this limited buy of turmeric proved to be very good and thus Nestlé decided to develop a long-term relationship with the family business. Today, Paras is one of the largest and most advanced spice processors in the entire country, with an annual production capacity of 12,000 t. It procures spices from all over the country: 40–50 % from the south, 20 % from the west and 10–15 % from the northwest.

Given the close relationship Nestlé has established with Paras, the spices producer and supplier has been influenced by the corporate practices the food company follows in many ways. For example, it has followed a very similar business model like Nestlé. The company does not own land or produce spices itself. Instead, these are bought from individual farmers, mostly smallholders, just like Nestlé procures its milk from independent producers. Also, like Nestlé, Paras gives technical assistance to the growers it works with. Paras equally gives very serious attention to the quality of spices it procures, since ultimately the Nestlé factories in Moga and elsewhere use much of this supply. Paras has thereby helped streamline an otherwise fragmented and geographically scattered spices supply chain of heterogeneous quality. Currently, around 5,000 farmers grow chilli, and 35,000–40,000 more are involved in the entire spice procurement process of Paras.

In 2007, responding to a mounting demand for chicory, Paras decided to experiment with its farming in 2 acres of land. This attempt was successful, and, accordingly, the company decided to produce the crop on a commercial basis from 2008. By 2011, some 400 farmers were growing chicory in around 500 acres. Paras sold over 1,000 t to Nestlé in 2011, including the company's factories in South Africa.

This successful partnership has expanded over the years to cover several new activities for Paras. Nestlé has recently introduced it to the cattle-feed industry. Paras is now constructing a plant for producing feed where the owning family will employ some 100 workers and is expecting to produce 200 t of feed per day. Paras will procure agro-commodities for this factory from some 8,000–10,000 farmers. The company is also constructing a starch-drying unit (employing 30–40 workers), at the direct suggestion of Nestlé. This activity is currently being carried out by Nestlé itself but is now facing capacity restraints.

Because of consistent good management records, the former mom-and-pop grocery store owner has now become a successful mini-conglomerate. It currently plans to expand to even more commercial activities, which will generate additional employment opportunities. None of these positive and synergistic developments would have occurred had the Nestlé factory at Moga not existed.

Paras is further planning to expand its spice production by almost doubling its capacity, employing additionally 100 more people. It is also establishing another factory at Pantnagar, Uttarakhand, where Nestlé opened a manufacturing facility in 2006. At present 90 % of the employees of Paras come from within 30 km radius of Moga, which has further generated additional employment in the region. Moreover, and as a part of the mutually beneficial and symbiotic relationship that the two companies started three decades ago, Paras' expansion ventures are being supported with Nestlé's know-how, experience in factory design and laboratory requirements, technical assistance, financial support, and the company's assured market for the outputs procured by Paras. In 2011, Nestlé accounted for nearly 65 % of the business activities of Paras.

7.4.3 Transportation Sector

One good example in this sector is Brar Enterprises Limited, a company that has been transporting milk for Nestlé since the factory was first established. The company has upgraded, improved and adopted its equipment and technology to respond to Nestlé's requirements, expectations and standards with in-house capital and increasing expertise. This has raised business practices quality setting of Brar Enterprises to a much higher level than its competitors. These positive steps were instrumental in opening doors for additional work with local and multinational companies who were attracted to Brar, primarily because of its reputation for providing high quality and reliable service. Currently, as a result of expanding opportunities and increased competition in the transportation sector, Nestlé is no longer its sole customer but still accounts for nearly 20 % of Brar's business activities.

With the encouragement of Nestlé, Brar Enterprises has expanded to new areas like packing sachets, dairy creamers and pasta for the company. Nestlé, in turn, has invested in the necessary machinery that is located in Brar's compound, which is used according to corporate targets and exclusively for its own products. Nestlé at present maintains an office within Brar Enterprises to ensure its stringent quality requirements are met consistently. As competing Government-sponsored cooperatives are likely to set up similar businesses in Moga, Brar is considering to provide comparable services to Nestlé's future competitors. With this objective, Brar is currently planning to purchase its own packing machinery which will give it flexibility to explore other new possible ventures with the other commercial competitors of Nestlé. This plan, if successful, will have the additional advantage of guaranteeing full-time work for its employees and thus a good and steady source of income for them.

7.4.4 Cattle-Feed Suppliers

Before the arrival of Nestlé in Moga, the supply of cattle-feed was unknown as an organized business activity. Considering the very low yields of buffalo and milk cows, Nestlé developed a formula for feed concentrates to provide a more balanced diet for the cattle and dairy animals. These feed are being sold to the Nestlé farmers on a non-profit making basis. Concurrently, Nestlé is encouraging better rotation of traditional fodder crops and also growing of lucerne as cattle-feed. Progressively, and as farmers have more financial resources to invest in their fields, more productive manual and mechanized farming technology are being introduced. This has been a continuing process. For example, in 1973, Nestlé assisted in the Government of India campaign to introduce the scythe instead of the sickle in harvesting fodder crops to improve production efficiency.

The veterinary experts of the company and its dairy agents on the milk roads trained farmers on the use and advantages of the new feed. With the support of veterinary staff and truck drivers, Nestlé has ensured a regular supply chain of this product to provide to the farmers. Sales of this specialized feed have increased from 1,500 t in 1973 to 8,939 t in 1991 in the Moga milk district. This has encouraged many new entrepreneurs to enter the cattle-feed supply business. This has become a new business which has provided them with additional opportunities to engage in formal and profitable economic activities.

In Moga two factories at present are making concentrated feed for cattle based on formulation supplied by Nestlé. One of them is P. Marka Registered, a company that has been historically engaged in mustard oil production from 1933. It decided to enter into the cattle-fed business in 1988 because of the explicit encouragement of Nestlé. Initially their plant had a capacity of 50 t/day, which was subsequently doubled to 100 t/day. However, with the still continuing and mandatory power cuts in the area, the plant is able to produce only about 75 t/day at present and has to forgo 25 % of its production capacity. These power cuts also affect Nestlé and all other ancillary firms in the region.

The company sells cattle-feed to Nestlé and in the general market. It produces and sales two differentiated products. Nestlé has provided Marka with a very specific formula to manufacture the feed it buys from the company. Accordingly, the plant has two separate production lines, discriminating between products manufactured exclusively for Nestlé and those of inferior quality sold at the local market at a cheaper price. As a buyer, Nestlé keeps a strict control over the quality of the cattle-feed and all its inputs and outputs. These benchmarks are considerably higher than those locally followed. Marka thus needs to comply with stringent quality requirements to maintain this business partnership. These efforts are supported by Nestlé through technical advice on how to meet, maintain and improve the higher standards. The company currently employs 25 people, only for Nestlé-related operations. Special training is given to each worker so that they can comply with Nestlé's strict quality specifications and requirements. Samples are regularly taken and then analysed to ensure consistent quality. A Nestlé representative visits the factory on a daily basis for quality control purposes. In addition, unlike other suppliers, Nestlé is particularly concerned with aflatoxins level of the feed. This is because cattle that may have been fed with contaminated feed could produce milk that could a health hazard for humans.

By following such standards, Marka has steadily gone up the quality ladder and the quality of locally sold feed has increased even if they still remains below the much stricter compliance requirements applied to products sold to Nestlé. This seems to be an issue related to profitability. The owner of the company explained that cattle-feed is very expensive not only because of the processing cost but also due to the price of raw materials. She also noted that even though the profit margin is low for Nestlé cattle-feed, Marka continues to manufacture it because of the other benefits it can reap such as enhanced reputation and assured regular payments. Associated capital certainly helps the cattle-feed producer attract more buyers for its products and predictable income helps smoothen the company's cash flow in an otherwise somewhat disorganized market. Such status-by-proxy is oftentimes related to the quality, hygiene and cleanliness spillovers the Moga factory has spread to farmers and ancillary firms alike.

7.5 Climbing the Quality Ladder

Nestlé has triggered positive spillovers in the region through its strict focus on hygiene, cleanliness, as well as quality controls and monitoring. This has directly and indirectly led local labourers, entrepreneurs and ancillary firms to 'climb the quality ladder' steadily. These quality upgrades have been used as benchmarks by consumers and have become a common denominator for other dairy stakeholders to follow. It has set a precedent and easily recognizable quality-price comparison and reference point for consumers and competitors in the region to consider. The company has successfully put in place a culture of improvement through rigorous quality control.

Working with Nestlé as an important and steady supplier has brought about substantial quality improvements in the mind-sets and actual working practices of all the associated companies in and around Moga. For instance, in order to do business with the Moga factory, Paras had to develop, implement and consistently maintain a stringent quality control system. Such strict standards have helped nurture the current strong relationship between the two companies. The fact that Paras has supplied Nestlé with spices for some 30 years has consolidated its credibility as an excellent and reliable spices supplier in India and has allowed the company to reap significant reputational gains from this well-established partnership. Thus, when other major multinational food-related company like Pepsico and Del Monte wanted to source spices, they came to Paras because of its well-established reputation as reliable supplier of good quality spices. Currently, Paras is expanding its spices operation, selling more and more of its products to other multinational corporations, both in volume terms and also as a percentage of its total operations.

Nestlé has not only been a reliable and long-term sustainable customer of many of its suppliers, but it is also one known for having very strict quality requirements and ethical business practices. It conducts a number of training programmes and continuously monitors a set of parameters related to the employment policies followed by its suppliers in terms of wages, employee health, safety and security, good manufacturing practices, hygiene standards, and the prohibition on the use of child labour. Collaborating firms have now successfully absorbed such responsible requirements and corporate behaviour. This has undoubtedly enhanced their reputation and public profile.

Similar to other ancillary factory employees, agents and dairy producers working with and for Nestlé, Paras has benefited from the reputational capital and social prestige associated with the brand. Nestlé has been one of the most important factors which has helped Paras in procuring contracts and deals with other important companies. It has given the spices company its 'seal of approval', which has automatically put Paras in the 'high quality' bracket in the Indian market.

More importantly, the relationships factory employees, agents, dairy producers and workers at the ancillary companies have established with Nestlé have surrounded them with brand-associated social prestige and trustworthiness. This reputational capital has been built from the interplay of organizational, economic, physical, human and technological resources Nestlé invests in and requires from the companies and the people it collaborates with. The resulting reliability, and attributed quality-by-proxy seems to have been particularly true for the ever-expanding array of service industries that have sprung up around the company's activities, production lines and the overall ecosystem. Numerous ancillary enterprises have developed a symbiotic relationship with Nestlé, tying most of their successes and business expansion plans to that of its business partner. Working and collaborating with the company has also had an aspirational impact on the associated companies and also in the community.

7.6 Chain of Other Nestlé-Related Activities

Electrical and mechanical workshops, as well as small factories doing minor repair jobs needed for the Moga plant, have steadily grown around it. In 1992, around 77 workshops were delivering electric and mechanical support to the factory. Building and construction businesses have also thrived because of Nestlé's presence and the needs of its suppliers. As the activities in the Moga factory have intensified, so have the requirements of its supply chain. These requirements have ensured that support industries like hoteliers, builders, craftsmen and entrepreneurs have prospered in Moga. Quasi-continuous extension of factory building since 1980 has ensured around 350 jobs per year for various national construction-related companies.

Nestlé's Moga factory has also supported the printing and packaging industry in the region. For example, in 1992, the 58,000 t of finished products needed cardboard containers, tin cans, sachets, labels, bottles and other packaging items. All these requirements have ensured that its ancillary support industries are growing and are generating more and more jobs. Nestlé's needs and demands have increased exponentially over the past decades. These have had positive impacts on an equally explosive growth of the ancillary support industries. In and around Moga, the impacts of this virtuous cycle are evident.

Hoteliers, builders and various other craftsmen and entrepreneurs have now settled in Moga to seize the business opportunities that have followed with the successful and continuous development of the region. This has helped diversify the community's economic base by creating a wide range of remunerated opportunities that go beyond traditional farm-related activities and have drawn in members from different demographic groups into the local labour force. These factors have contributed to discourage outmigration from rural settlements to urban centres by enlarging the employment possibility frontier.

Reference

Deep G (2002) The octroi debate: the case of Punjab. In: Singh UB (ed) Revitalised urban administration in India: strategies and experiences. Kalpaz Publications, New Delhi, pp 265–275

Chapter 8
A Test to the Moga Community of Interests

Over the years, the consistent transparent, predictable and reliable business case Nestlé has followed in Moga has made the villagers trust the company. It has also made many of them loyal suppliers. As one farmer who was interviewed for this study said, "loyalty and trust have to be a two ways street". Trust lies at the core of the community of interests Nestlé has established with Moga's farmers. This trust and staunchness in the company can be best indicated by considering what transpired during the decade-long period of terrorism in Punjab, from 1982 to 1993. During those years, the separatist group Khalistan Commando Force proclaimed the independence of Punjab and engaged in a series of terrorist activities to attain its goal.

8.1 Trust in Times of Crisis

Terrorism initially started in the districts of Amritsar and Gurudaspur but soon spread to Hoshiarpur, Jalandar, Ludhiana and Faridkot. At that time as one of the more industrially developed area in Punjab, Moga was part of the Faridkot district until it became the state's 17th district in 1995. The region quickly became one of the worst terrorist-affected areas. By a conservative estimate, some 21,469 people were killed. When the police and army managed to control the terrorist activities being carried out in the urban areas, the villages and towns of Punjab faced the main wrath of the terrorists.

As the terrorists moved to the villages and the security forces followed them there, they brought about many hardships to the local population. Furthermore, rival gangs within the terrorist groups fought each other, with innocent villagers often caught in the middle. As the problems intensified, population migrated massively from rural to urban areas to escape the violence of terrorism. Labourers working in Moga from outside Punjab returned to their hometowns because they were afraid for their lives. These two population displacements had a negative impact on the economic activities of the region due to extensive labour shortages. Unsurprisingly, during the terrorism years, agriculture, the state and region's most

A. K. Biswas et al., *Creating Shared Value*, SpringerBriefs on Case Studies of Sustainable Development, DOI: 10.1007/978-3-319-01463-0_8, important economic activity, was severely affected. Growth rates in the sector plummeted from 6 to 2 %, curtailing the farmers' investment, migration and social capital decisions (Singh 2008).

It is truly noteworthy that even though the farmers of Moga and the surrounding area were going through a difficult and challenging time of widespread fear, mutual suspicion, lawlessness and violence, the functioning of the Nestlé factory and its milk collection activities continued for the most part as normal. This can be seen in terms of total milk procurements during these turbulent years and also from the number of people who continued to supply milk to the company (Table 8.1). On average, from 1980 to 1996, around 46,937 farmers supplied 98,563 million kg of milk. Procurement, in terms of volumes of milk and farmers delivering it, peaked in 1990 due to company's efforts to reach out to farmers in remote areas.

It is worth pointing out how the overall trends during this difficult period were of increasing milk supply as well as of growing number of suppliers. These are shown in Table 8.1. This does not only bear witness to the extent dairy activities had been entrenched in the economics of the region and household livelihoods but also how the trust farmers had placed on the company continued unabated. In fact, perhaps it was further strengthened even when their security could not always be guaranteed. After a few decades of having set up the Moga factory, the dairying sector already played such an important role in the economic life of the community that farmers went to great lengths to maintain and increase, whenever possible, their milk income.

Table 8.1 Milk collection for Moga factory during terrorism years

Year	Average milk price per kg of fat (in INR)	Total yearly milk intake from own sources (in million kg or thousand t)	Number of milk suppliers (thousands)
1980	31.6	55	32
1981	34.0	35	32
1982	39.3	58	34
1983	42.0	72	35
1984	45.5	71	37
1985	51.0	114	50
1986	51.3	117	52
1987	52.8	107	52
1988	62.3	98	42
1989	71.1	127	52
1990	71.6	136	58
1991	83.8	100	50
1992	103.1	103	46
1993	95.9	106	41
1994	105.1	85	42
1995	138.3	84	41
1996	141.7	109	55

Source Nestlé Records

In those years, milk collected and the number of suppliers declined only in 1988. Our analysis indicates that the main cause for this decline was not terrorism but the vagaries of nature. The summer of 1988 was very hot and the monsoon arrived late. However, the main cause for much of the decline could probably be attributed to the torrential rains in the upper reaches of the Himalayas, which contributed to heavy floods in the plains of Punjab. The floods affected some 1,453 villages and inflicted serious damages to houses, crops, roads and embankments. Schools and colleges had to be closed so that flood-affected families could take shelter in them. Naturally, milk collection was adversely impacted during this natural calamity even when the Agricultural Extension Services offered by Nestlé were defining in mitigating the consequent losses.

Another important factor that could explain this decline in milk collected and number of suppliers was the lack of seasonal labourers, who mostly came from the state of Bihar. Frightened by the increasing insecurity because of terrorist activities, most workers decided to return to the safety of their hometowns. From May 1987 to April 1988, terrorist activities increased very significantly in Punjab resulting in 1,533 casualties, including 109 policemen. Consequently, the Central Government increased anti-terrorist efforts to dismantle and demobilize militants from the Khalistan independence movement. The exodus of hired labourers in masses from farms to their home towns affected milk production and agricultural activities significantly, resulting in the decline of agricultural outputs and also in the number of milk suppliers. This trend, nevertheless, was soon reversed and milk volumes went back to their previous levels and even surpassed them as the terrorist activities were brought under control.

Considering the context and situation prevailing at that time, it is indeed remarkable that Nestlé maintained its milk collection system during the years of violence, and even saw some improvements in its procurement process. Interviews with the farmers of the Moga district indicated that Nestlé milk trucks continued to arrive at the collection centres almost on time every day, even though all other forms of transportation at that time were severely curtailed. Over those very difficult years, the Moga factory was closed only for two days, and this was primarily because of the general strike that was organized by the ruling Congress Party following the assassination of the then Chief Minister of Punjab, Beant Singh, in 1992.

Several farmers also referred to the situation in Moga during the India-Pakistan war of 1971, especially in view of the fact that Moga lies fairly close to the border between the two countries. The movement of vehicles during the day was restricted and blackouts were strictly enforced at night but Nestlé's milk collection trucks arrived regularly early morning every day. The company proved it was a predictable, reliable and resilient business partner that did its best to continue with its normal factory activities despite the adverse political context under a very difficult situation. Even more so during those challenging years, this committed and responsive corporate behaviour proved particularly reassuring to the farmers and their families. Not surprisingly, they responded in a similar fashion.

The farmers interviewed were categorical in their views that the respect and confidence Nestlé had earned over the decades enabled the company to continue its milk collection activities without any serious setbacks. This was only possible because of the fair, transparent and positive rapport Nestlé had consistently established and nurtured with some 50,000 families of the area and the significant economic benefits accruing to the farmers at that time. This relationship is based mostly in terms of trust and mutual self-interest and spur from Nestlé's ethical behaviour, transparent business practices, fair and timely payment for milk purchases, agricultural and veterinary support services, and contributions to the region's economic prosperity and social development. The local civil administration, which acknowledged and appreciated the benefits the Moga factory had brought to the region, went out of its way to ensure that milk collection continued without major disruptions despite the serious terrorism threat in the area.

Initially, the people in the region were afraid that the prosperity they had carefully nurtured over some two decades of hard work might not continue due to the war or because of terrorist activities. They thus tried to return the support they had received from Nestlé so that the Moga factory could continue to survive and thrive. This once again reinforced the important linkages and interdependence between the company's long-term sustainability and the farmers' wellbeing, and vice versa.

8.2 The Long Process of Building Trust

All the people that were interviewed for this study (excluding factory employees) gave a consistent narrative of how had the establishment of the Moga factory positively changed the lives of the people in the region and the region itself. Many of these advantages came about as the company consistently seized local potentials and turned them into long-term opportunities for the company as well as for its suppliers.

At first, the study team noticed a generational change in the loyalty experienced and younger farmers expressed towards Nestlé. Initially, it was concluded that new generation farmers do not factor in technical assistance in making the decision as to which company they sell their milk to and that they are being attracted by the offered price alone. Yet, on a closer look, their commercial behaviour and preferences could also reflect a radical attitudinal and generational change regarding business practices. This is a contribution to the region's corporate culture Nestlé can take credit for establishing and nourishing.

Younger farmers have grown in an area where, for many decades, the corporate dairying culture has meant the purchasing company looks after their interests and pursues a wide range of social goals. Farmer prosperity and company success are seen as interdependent and mutually reinforcing. This is has become the de facto process through which milk business is conducted in the region. As such, these expectations could also be taken as an indication of the extent to which a new milk and corporate culture has taken hold in the area. This change is a promising and an important area of study that is yet to be explored, especially given the attitude exhibited by new generation. They are fully aware that if a company, Nestlé or any other of its competitors, is to reap competitive benefits, these ought to be tied directly to the community's wellbeing.

For many farmers, the interdependence that exists between a thriving community and a successful company was not all that evident. Coming from times when this was almost unthinkable, this business culture or economic behaviour is something that elder farmers would not take for granted. They proudly point to the 50-year very positive business experience they share with Nestlé and are always aware of the transactional and relational benefits this partnership has delivered to both parties. This older group has continued to favour milk supplies to Nestlé due to: steady and timely payments and the convenience in which they are made; accurate, strict, fair and transparent quality assurance controls and measurements; supporting services and training; the trustworthiness and responsiveness of accountability mechanisms; corporate reliability and reputation; and expectations of positive technological, managerial, technical and skill spillovers. They also recall the social, economic and environmental conditions that prevailed in the villages when Nestlé first arrived, and how these have progressively improved over time, primarily due to its support.

With time, notions of company-society interdependence, what Nestlé now calls 'Creating Shared Value', are being more widely internalized by younger and more experienced farmers alike. Albeit farmers' trust in and loyalty to Nestlé and intensified local competition have led them to be growingly aware of the high demand for their products, especially milk. With new players coming into the region, farmers now have the opportunity to seek higher prices for their goods and to use their production and management skills in similar industries. Consequently, they are all becoming more and more open to work with other buyers if they are offered more competitive milk rates and at least a similar level of production support system and technical services.

Reference

Singh P (2008) Causes and consequences of terrorism in Punjab: a rationalist perspective. Institute for Defence Studies and Analyses. Available via DIALOG, New Delhi. http://www.idsa.in/idsastrategiccommentsCausesandconsequencesofTerrorisminPunjab_PSingh_180708

Chapter 9
Evolution of the Dairy Industry in Moga

In retrospect, Nestlé's activities in Moga initiated the beginning of the dairy revolution in Punjab in 1961. It was the only milk procurer in the Moga milkshed area for years. It developed, prepared and nurtured the growth of the dairy industry in the entire region. This was accomplished despite the fact that for the first two decades, pricing was Government-controlled and depended on cooperatives. In fact, because of Nestlé's hard work over the past five decades, Punjab in 2012 contributed to about 10.5 % of total dairy production at the national level despite having approximately 3 % of the Indian dairy population. This is an important nutritional contribution to the country, especially as milk tends to be the only acceptable source of animal protein for vegetarians and many times the only intake landless, small and marginal farmers can afford.

When Nestlé established the factory in Moga, the company had realized the roles that the dairy industry could play for the overall economic and social development of rural Punjab. Even today, about 75 % of the population in Punjab live in rural areas and are engaged in agriculture to earn their livelihood. When the Moga factory was established, this percentage was significantly higher.

In 1982, Nestlé was buying 21.5 % of total milk produced in the Moga milk district, a figure that reached 24.7 % by 1983. Around this period, milk procured by Government cooperatives or other private companies was insignificant, as most of them were establishing plants in other areas of Punjab, but not in the Moga district. As noted earlier, the success of Nestlé in Moga has attracted other dairy industries to the area. Factories belonging to the Punjab State Cooperative Milk Producer's Federation Limited (Milkfed) in Amritsar, and Glaxo Smith Kline at Nabha are only a few examples. Milkfed started operations in 1973 and has now become an important milk cooperative in Punjab.

In 1991, competitors started to establish plants even in Moga. By 1992, Government cooperatives were buying 5.9 % of total milk produced in the district, with other private companies buying 3.1 %. In those years, the share of Nestlé remained almost constant at about 22.5 %. Between 1982 and 1992, total milk production increased in the Moga milk district from 265,000 to 467,000 t. In 1992, competing firms were operating in 192 village dairies using Nestlé's milk procurement model. That year, they bought around 27,000 t of fresh milk, equivalent

A. K. Biswas et al., *Creating Shared Value*, SpringerBriefs on Case Studies of Sustainable Development, DOI: 10.1007/978-3-319-01463-0_9, © The Author(s) 2014

nearly to one-quarter of the amount delivered to Nestlé (Nestlé 1993). The arrival of competitors cost the company approximately 9 % of its milk intake in 1992. Nonetheless, from a development perspective, this is a testimony of vital impetus given to the region by Nestlé and one that has significantly benefited the farmers in Moga and all the consumers of milk products all over the country.

By 2002–2003, in addition to Nestlé, there were 11 Milkfed dairy plants and around 30 other owned by the private sector. Milkfed was procuring around 1,525,000 l/day and private sector companies were procuring 3,255,000 l/day. If today, the total input of Indian dairy sector exceeds the combined output of paddy and wheat, one of the major factors is rapid development of the dairy industry where Nestlé has played a very important role.

There is no question that groundwork carried out diligently by Nestlé's Agricultural Extension Services for three decades has also benefited rival firms. The company has undertaken considerable time, effort and monetary investments in building, strengthening and fostering the human capital, skills and capacities of the milk producers and oftentimes of their families at large, ancillary firms, and other suppliers it works with. As the pool of experienced farmers and milk producers has expanded, so has the attractiveness of the region for other firms interested in carrying out similar commercial activities to start operating there. When arriving to Moga, competitors can reap the long-term benefits of knowledge, quality, reputational and expertise positive spillovers and externalities Nestlé has successfully brought to the region. These new competitors are fully aware of the region's solid milk culture, the sturdy sector infrastructure and are counting on being able to engage a portion of the skilled farmers and milk producers of Moga.

Reference

Nestlé (1993) Nestlé in India 1962–1992. Nestlé, Vevey

Chapter 10
Moga as a Catalyst for Development

During the 1960s, 1970s and 1980s, considerable benefits that accrued to the region can be for the most part traced back to the presence and activities of Nestlé. However, as the Indian economy expanded during the post-1990 period, the levels and extent of its impacts, though still significant, are not of similar orders of magnitudes as the changes that were observed during the first 30 years of the Moga factory. By then, national and state-level macroeconomic growth led to an overall improvement in the communities' physical capital and available infrastructure. Villagers began to be connected to the electricity grid; the use and ownership of landline and mobile phones exploded; accessibility by land to the region improved and market access to and from Moga was facilitated as the transportation system was modernized, the number of paved roads grew and the number of motorized vehicles increased steadily; and irrigation systems and fodder-chopping activities were mechanized.

The overall increasing prosperity in the region that followed growth inducing Central and state level Government policies was also a positive contributory force. Based on our own analyses and extensive interviews with 221 people in and around Moga, there is absolutely no doubt that Nestlé has played very important direct and catalytic roles to improve significantly the social and economic conditions and knowledge levels of the area from the early 1960s. As such, the combination of public and private efforts has consistently worked together to make of Moga a progressively thriving community.

Broadly speaking, with better-remunerated and expanding employment opportunities, the region's economic prosperity also inspired farmers to have their children attend school and receive formal education, especially girls. Shortly after the Moga factory became operational, an elementary school opened in Dudhike village, which was later upgraded to a high school in 1966–1967. The first graduate college in Moga opened in 1968 and numerous female students enrolled. Soon, English-medium schools started to mushroom in the area. The impacts of higher education, entrepreneurial awareness and desire to lead a more mobile life subsequently encouraged many families to immigrate to other countries. In Dudhike village alone, at least one hundred families (nearly half) have settled abroad, primary in Canada.

A. K. Biswas et al., *Creating Shared Value*, SpringerBriefs on Case Studies of Sustainable Development, DOI: 10.1007/978-3-319-01463-0_10, © The Author(s) 2014

Though the company is still continuing with its development-related support activities, these were very evident and of particular importance during the first two decades of the Moga factory. During the post-1980 period, various levels of Government started to gradually give more emphasis to certain types of community-building activities. Such schemes have been part of the company's long-term commitment to rural development, reduction of rural poverty, development of the agricultural sector and promotion of rural non-farm related economic activities. Even after this improvement, what is being undertaken now can meet only the needs of some of the people for some of the time. In addition, the extent and the quality of services delivered by the Government institutions in the rural areas around Moga, as well as the rest of the country, still leave much to be desired. Much more work remains to be done, and the private sector companies like Nestlé will continue to play important roles in the future.

Even though the concepts of Creating Shared Value and Corporate Social Responsibility are of comparatively recent origin, the founders of the Moga factory realized, as early as 1961, that if the company was to succeed and thrive, they had to provide many of the social services needed, which very few, if any, public or private sector companies in rural Punjab were delivering at that time. They realized that if their suppliers were to be encouraged to assure the development of a milk economy, Nestlé not only should pay the farmers a fair price for the product, but would also have to walk the extra mile to improve the economic, social and environmental conditions under which they lived. They were fully aware that the standard of living as well as the quality of life of the farmers had to be substantially improved so that an increasing supply of good quality milk could be produced and then procured. This understanding was absolutely critical for the company to survive and to grow. In other words, the successes of the farmers and Nestlé were strongly intertwined, and Nestlé realized this fact from the very beginning.

Very few corporations, if any, were following at that time profitable business strategies that also were socially beneficial. Only in recent years, the concepts of Creating Shared Value and Corporate Social Responsibility have been coined, developed, adopted and implemented to a certain extent. It is very evident from the current study that the pioneers of the Moga factory were not aware of the Shared Value Philosophy, or Corporate Social Responsibility, very much like the rest of the world, but they quickly realized its relevance and importance. They understood that unless the social and economic conditions of the farmers could be significantly enhanced, the factory at Moga had no long-term future. They thus introduced a series of actions which would raise the standard of living and quality of life of the farmers. This would ensure that the region would develop a sustainable milk economy. If and when economic and social conditions of the farmers could be substantially improved, an increasing supply of milk could be assured. This, in turn, would guarantee the success of the factory. They appreciated that the arrangement had to be a 'win–win' situation for both the farmers and Nestlé and that issues like economic efficiency, company success and corporate growth were indivisible and inseparable from community progress and welfare. They further concluded that farmers had to 'win' first so that the company could obtain its raw material and thus win thereafter.

The Moga pioneers thus arrived not only at a philosophy of long-term and sustained change and a corporate culture of improvement unknown 50 years ago. They actually implemented it successfully. They realized that, even though Nestlé is not a development agency or a philanthropic institution, for it to survive and prosper, the company must look after the welfare of its suppliers and the Moga region as a whole. This needs to go well beyond simply procuring milk at a fair price. Far from being perceived as prospering at the expense of the community, the company established a new ethos that fostered active but historically ignored partnerships.

The planners of the Moga factory further realized that the fate of the factory and the fortunes of the milk suppliers were intricately interlinked. If the farmers succeeded, so would Nestlé, and vice versa. Thus, they went out of their way to try to provide the means to ensure the success of the farmers. This shows the mutually reinforcing and beneficial interdependence between a company and the community where it operates at its best. They did not talk about it, articulate it explicitly or wrote about it for public relations purposes: they simply went ahead and did it.

Such rationale involved completely rethinking the existing business models, as well as reordering and streamlining the then accepted wisdom of supply chain relationships, not only for when the factory was started but also for the future. They put in place innovative, sustainable and forward-looking business models, which by any standard were far ahead of their time. They also imposed on them higher social, economic and environmental standards that have been further improved regularly over the decades. In the process, and by deliberately pursuing rural development initiatives as part of its corporate model, they set in motion a process that has lifted well over a million people from crushing poverty directly through their own activities and also indirectly through their suppliers.

Annex 2 shows how different development indicators have significantly progressed over the last 50 years in Moga, as well as Nestlé's roles in catalyzing these changes. Whilst many other factors like the Green Revolution and evolution of appropriate policies at different levels of Government have helped, it should be noted that without the presence of Nestlé at Moga, many of these improvements would not have taken place within this time period and the magnitude and distribution of the beneficiaries would most certainly have been significantly less. The sequence of these changes is shown in Annex 1 as timeline diagrams.

10.1 Creating Shared Value

Nestlé has quickly responded to immediate and long-term needs that strengthen the links between the company and the community within which it operates, oftentimes through schemes that partly fall under the current common definition of Corporate Social Responsibility. Many private sector firms have begun to follow this path.

However, Nestlé's Creating Shared Value programme, the goals of which have been implemented over the decades at Moga, transcends well beyond what is considered to be Corporate Social Responsibility at present. For example, these initiatives have covered a wide range of immediate needs as well as longer-term and sustainable community development programmes. Addressed issues have included, but not necessarily limited to, nutrition, hygiene and sanitation, humanitarian assistance at time of natural disasters,[1] and improved access to clean drinking water. Additionally, Nestlé has provided multiple social and supporting services that have included support to and participation in immunization campaigns, arranging medical camps and providing basic credit services.

10.1.1 Improved Hygiene

The company's extensive awareness building efforts have been very useful in making milk suppliers realize the importance of maintaining good heath, hygiene and environmental conditions at both household and farms levels. These efforts have encouraged and facilitated behavioural changes leading to better hygienic and nutritional practices and a reduction in the prevalence of various diseases. For example, farmers are now aware of the potential impacts of pesticide and fertilizer residues in milk. This understanding has helped farmers to reduce the levels of pesticide they use. It is also reflected by the fact that despite malaria being endemic in the region, use of DDT has been considerably reduced in the milk collection area.

Many villagers stated categorically that it was the company's emphasis on cleanliness and quality, starting by the factory premises and then visits of Nestlé officers to their homes that made them aware of the importance of keeping their homes clean and adopting good living habits. Despite local conditions, practices and culture, Nestlé has created new, much higher, standards for cleanliness and hygienic conditions in Moga. This has conclusively shown that there is absolutely no reason as to why such standards cannot be maintained and continually improved in similar settings of small rural villages or in places where infrastructure, roads and communications are limited.

Suppliers of inputs to Nestlé, whether they are small farmers or large producers, all have been inculcated with the importance of maintaining a consistent focus on quality and good environmental conditions. Continuing emphasis has made people conscious of quality and its importance on all aspects of life. Nestlé can thus be credited with contributing to significant improvements in family health over the years as it has promoted increased access to hygienic and nutritional information.

[1] For instance, during the 1988 floods that afflicted the region, Nestlé was at the forefront of relief operations, distributing food, fodder, and veterinary services to the stricken people.

10.1.2 Nutritional Benefits

The entire chain of Nestlé activities is focused on good health, and nutrition. On the manufacturing end, the company has consistently stressed the importance of a healthy environment, which in turn leads to healthy animals and thus good milk quality. This is then further reinforced by proper milk collection and transportation as well as practices at the factory to produce milk-related products. Such actions once again show how social and corporate goals have been consistently aligned and co-ordinated to produce good results for both the company and the people associated with it.

Some nutritional benefits have accrued directly to the community at Moga and are the result of deliberate efforts, campaigns and programmes carried out by Nestlé. The Nestlé Healthy Kids programme is one such scheme. Conducted in and around Moga, in collaboration with the Punjab Agricultural University, the initiative takes place in Government schools with the objective of creating and raising awareness on good nutritional practices, cooking methods to enhance the nutritional content of foods and physical fitness among students in village schools. It has already reached to about 5,500 girl students between the ages of 13 and 17 years by 2012. This same Healthy Kids programme is also being implemented in other regions in the country (Nestlé 1993; Nestlé India Limited 2000, 2010, 2011) where Nestlé has a presence.

Others advantages, however, have been more widely extended to consumers through nutritionally enhanced Nestlé products. The company was prompt to notice that many children in the region and the country were iodine deficient, which impairs their mental and physical development. It thus began to manufacture Maggi bouillon cubes using iodine-fortified salt and later on to add iodine to almost all of Maggi's product range so that the deficiency can be eradicated or reduced.

10.1.3 Filling in Local Financial Gaps

During the early days of the Moga factory, the company filled in many service gaps, including those related to banking and finance. It thus extended credit and allowances for various production-enhancing purposes by putting in place affiliation incentives for milk suppliers and deducting loan repayments from milk payments. From the very onset of its operations, the Moga factory applied for a loan for INR 1.5 million so that it could extend credit to farmers interested in purchasing more buffaloes and thus increase productivity. Nestlé was to repay the loan over a period of 4 years at 4.5 % interest rate. These were the exact same conditions it offered to the borrowing milk producers. Farmers repaid the loans by having a certain amount deducted from milk payments every fortnight, except during the lean months of April, May and June. Another such example dates back to 1971, when Nestlé supported a local bank's loan for INR 200,000 to farmers in the region to purchase more dairy animals. In total, from 1971 to 1979 loans of this kind to the farmers totalled INR 11.275 million (Nestlé 1993; Food Specialities Limited 1968, 1971).

Nestlé discontinued these services after formal banking institutions began delivering them to the farmers in the region. It has steadfastly encouraged the local banks to extent credit lines for dairying related activities and inputs. The company currently assists credit-seeking farmers to get loans from the local banks by acting as a form of employment, income stability and reputational collateral. From 1980, local banks have begun to accept the farmer's milk pay as a repayment guarantee and collateral. This simply would not have happened without the strong support from Nestlé. That same year, local commercial banks extended credit for more than INR 5 million in loans to local farmers to finance expansion in the agricultural and dairy activities between 1980 and 1989, 2,436 farmers took advantage of such loans and bought over 4,300 dairy animals (Nestlé 1993). In addition to all the productivity gains in milk production because of steadily improving animal health conditions and Nestlé's technical services, access to credit allowed farmers to increase the size of their herds and thus the volume of milk supplied to the company. Higher milk sales automatically translated into higher incomes and the repayment of the acquired loans by the farmers. This also ensured that Nestlé had an increasing supply of inputs needed for its factory.

10.1.4 Village Water and Sanitation

During the mid-1960s, Nestlé conducted house to house surveys in one of the most prosperous villages of the area, Madoke. The survey found that since none of the hand pumps had a concrete foundation, all the contaminated effluents were percolating to the wells and reaching water that was subsequently used for drinking. Persuading villagers to construct concrete foundations for hand pumps proved to be a very difficult task. The vast majority of the village's buildings and houses were built with mud. Thus people wanted to use concrete for the foundations of their houses rather than for the hand pumps, which they felt would be a waste of concrete and money.

The close relationships farmers and Nestlé officers formed were instrumental to convince the villagers engaged to lay the concrete foundation for the water pumps first and then take care of the foundations of the houses. One such villager was Puran Singh, of the nearby village of Mangewala, who agreed to the company's suggestion simply out of trust and not because of conviction. When the villagers saw the actual health benefits of such improvements because of improved quality of available water, other people soon decided to follow the idea. Laying concrete foundations for hand pumps served to improve the overall long-term health conditions of the villages.

As mentioned in previous sections, in the early days, loans were distributed for tubewell installations and Nestlé's engineers helped villagers to procure tubewell equipment of the appropriate quality at concessional rates and also provided technical advice at no cost. In 1968, Nestlé began an accelerated irrigation programme consisting of drilling 357 tubewells and installing power driven pumps, which contributed to increases in wheat yields by about 50 %. This also meant that rice could be grown from July to October. These inputs were introduced well before the national efforts to intensify agricultural production and thus had perceptible economic and health impacts. It also helped the area to quickly reap the benefits of the Green Revolution and mechanized and modernized agriculture which came a little later.

In addition, the company noticed that important community health gains could be attained if village schools near Moga gained access to clean water. Accordingly, Nestlé started a clean drinking water programme for some Government schools from 1999. The company considers those schools for provision of clean water where sizable number of students would benefit, no good water source is available and the communities cannot afford to do so exclusively but are willing to bear 10 % of the cost. After identifying the schools, Nestlé arranges to drill deep wells (50–120 m) and an electric submersible pump is used to lift water. It is then stored in food-grade plastic tanks, enclosed in a specially designed concrete facility. The 8-sided facility has 8 taps so that several people from the school can concurrently drink water, and thus reducing waiting time to a minimum.

The schools where such water facilities are provided have to formulate committees where students are responsible for maintaining the facilities. Water samples are regularly taken and then tested to assure quality. Nestlé provides regular maintenance and repair. As of 2012, these types of water facilities have been constructed in 116 schools, benefitting over 40,000 students and teachers. These schemes are then reinforced through water awareness programmes teaching students the value of water, its quality and health implications, conservation and protection of water sources and stored water, as well as responsible consumption and use of water. Such educational programmes have reached 20,565 students.

Nestlé's limited endeavours in providing access to drinking water and improving healthy habits have borne fruits. Despite many constraints, by 1978, drinking water quality has improved and became acceptable in seven out of nine villages where surveys were carried out. The set-up of water fountains and tanks in Government schools has had such a positive impact that farmers often would request the company to extend similar service to the villages as well. In recent years some Government institutions have started to take an interest in village water supply. Some villages are now receiving attention but significantly more work remains to be undertaken before they all receive access to clean drinking water.

Another important change Nestlé initiated in the rural areas has been the promotion of biogas plants from animal wastes and introduction of smokeless stoves. This created a revolution in villages during the early 1960s, when mostly wood, cowdung or coal were used as cooking fuel. Thus, indoor air pollution was a serious health hazard. By 1987, Nestlé had distributed some 2,285 smokeless stoves in different villages. This initiative has significantly reduced indoor air pollution, consequently bringing about long-term health improvement for the farmers and their families. Such activities have fostered more desirable energy use practices at the household level. It also had the advantage of disposing animal wastes in an environmentally-friendly way. Even after all these efforts, a main hygienic concern remains, that is, how best to dispose of animal wastes in a socially and environmentally acceptable manner in all the villages.

Unfortunately, hygienic practices have remained somewhat poor in most of the villages. Absence of latrines in rural areas forces women to defecate in open fields. Identifying this social needs and infrastructural gap, Nestlé encouraged villagers to work with the company to design, construct and maintain dry pit latrines. Such efforts have taken place along public awareness programmes to enhance good sanitation and health practices.

These contributions to individual, household and community health were particularly relevant for the first 20 years after the plant opened its production lines, especially as the Government programmes in Moga, and other rural areas, were mostly conspicuous by their absence. Overall, sustained campaigns that the company has initiated continue to make positive contributions to the lives of many of the villagers despite the arrival of many Government-supported efforts in recent years.

All these programmes, initiatives and efforts are part of the company infrastructure that has helped to cement the interrelationships between the Moga factory and the community around it. This has, in turn, helped to build a sturdier and more resilient local social, economic and environmental landscape. These locally contextualized initiatives respond to community-specific needs whilst all throughout expressing the company's worldwide corporate interest in building a healthy, robust and profitable milk district. Economic and social values are simultaneously fostered as local incomes are safeguarded, poverty rates drop, agricultural capacities are built, the living conditions of farming communities are enhanced and natural resources are efficiently used.

10.1.5 Environmental Conservation

The volume and variety of products manufactured at the Moga factory has steadily increased during the past 50 years. Simultaneously, production processes have been regularly modernized as a result of which their efficiencies have steadfastly improved. Thus, many of the resources needed, like water or energy, to manufacture each tonne of products have continued to decline since 1997.

While detailed facts and figures for the Moga factory were not available water consumption to manufacture each tonne of product has declined by some 74.3 % between 1997 and 2010 for all the Nestlé factories in India as a whole. This has been possible by institutionalizing the importance of water conservation throughout the production process and factory management. Water audits have been carried out regularly to identify where and how best water use practices can be introduced. New and innovative technologies are consistently considered so that

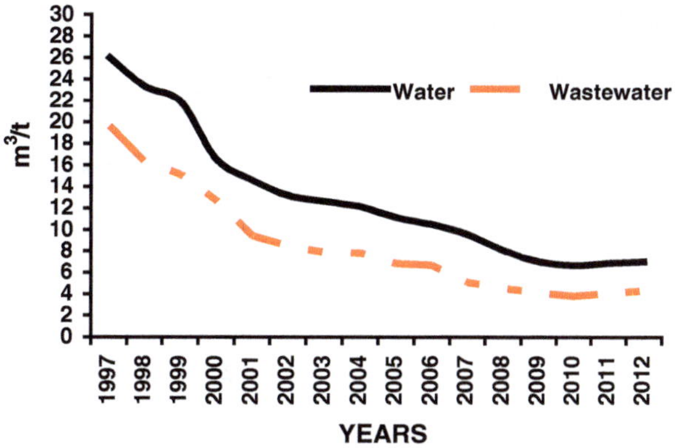

Fig. 10.1 Water consumption and wastewater production per tonne of product, 1997–2012. *Source* Nestlé records

improved water use efficiencies can result, even though the cost of water is a very minor part of the production processes. Equally, treated wastewater is being extensively reused wherever possible. The decline in water consumption per tonne of product during this period in the Nestlé factories in India is shown in Fig. 10.1. As water consumption has gone down, as a corollary, generation of wastewater has declined by nearly 73 % during the 1997–2012 period (as shown in Fig. 10.1).

A new prospect that is being considered is to capture water from milk that currently evaporates when powdered milk is produced. It is expected that a new process will be in place during the next two years to capture this evaporated water that has historically been lost to the atmosphere. The idea is to capture this water with the currently available technology, and then use it for the factory. When this occurs, it is highly likely that the Moga factory will become close to water neutral in its operation.

Like water, energy required to manufacture each per tonne of product has gone down by some 65 % during the 1997–2012 period in the 8 factories and manufacturing facilities Nestlé has established in India.[2] This significant reduction has been made possible by regular energy audits identifying where energy use efficiencies could be improved. Not only energy inefficient equipments have been replaced with more efficient ones but also special effort has been made in terms of new investments in installing new processes and process modifications to reduce energy consumption per tonne of product. In addition, waste heat from one process

[2] Nestlé India has presence across the country with 8 manufacturing and processing facilities, 4 branch offices and a head office in Gurgaon, Haryana. As already mentioned in this study, the company's first factory was established in Moga (Punjab) in 1961 and was followed in 1967 by a second plant in Choladi (Tamil Nadu). In 1989 a new factory was set up in Nanjangud (Karnataka). In the 1990s Nestlé opened up three more plants: one in Samalkha (Haryana) in 1993; and two in Goa at Ponda, in 1995, and Bicholim, two years later. The last two factories have been established at Pantnagar, Uttarakhand, in 2006 and Tahliwal, Himachal Pradesh, in 2012. The company also has 4 branch offices in Delhi, Mumbai, Chennai and Kolkata to handle sales and distribution activities.

Fig. 10.2 Energy requirement per tonne of product in GJ/tonne, 1997–2012. *Source* Nestlé records

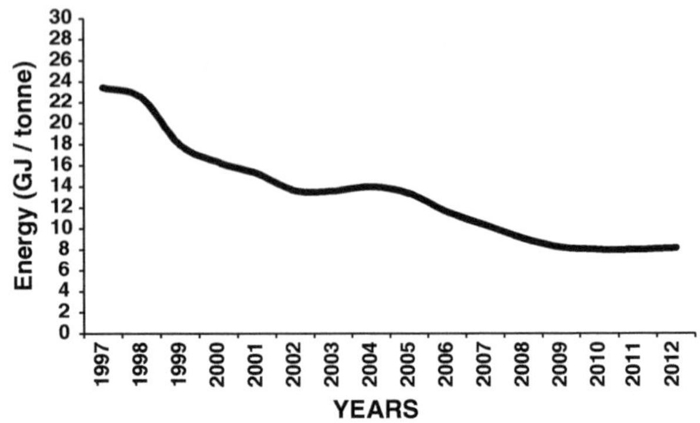

is being increasingly used as energy input for another process, thus improving energy use efficiencies of the factories. This has helped to contain total energy costs, especially as unit energy costs, unlike water, have steadily increased over the past decades. In addition to the enhancing environmental conditions, at the Indian plants of Nestlé, these steps have improved the economic efficiency of the plant by using increasingly less and less energy. The steady declining energy requirements to produce one tonne of product at the Indian plants of Nestlé between 1997 and 2012 are shown in Fig. 10.2.

Improving energy efficiency by using state-of-the-art technologies to improve fuel combustions have also contributed to more than 2/3rd reduction in carbon dioxide emissions (CO_2). This also proves that the steps taken to reduce industrial environmental and energy footprints can concurrently improve the economic performance of the factory. The steady reduction in CO_2 emissions per tonne of product, between 1997 and 2012 is shown in Fig. 10.3.

Fig. 10.3 CO_2 emissions per tonne of product (tonne/tonne), 1997–2012. *Source* Nestlé records

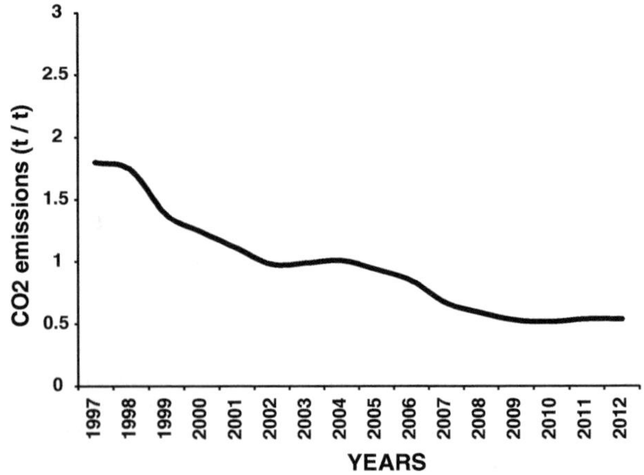

Consistent with the Nestlé model in Moga, the company knows that environmental conservation and the more efficient use of natural resources is a shared responsibility. Playing its part as a development catalyst in the region, it has made a commitment to environmentally sustainable business practices.

References

Food Specialities Limited (1968) Report and Accounts for the year ended 31st December, 1967, Directors' and Auditors' Report, New Delhi

Food Specialities Limited (1971) Report and Accounts for the year ended 31st December, 1970, Directors' and Auditors' Report, New Delhi

Nestlé (1993) Nestlé in India 1962–1992. Nestlé, Vevey

Nestlé India Limited (2000) 41st Annual Report, 1999, Directors' and Auditors' Report, New Delhi

Nestlé India Limited (2010) Annual Report, 2009, Directors' and Auditors' Report, Gurgaon

Nestlé India Limited (2011) Annual Report, 2010, Directors' and Auditors' Report, Gurgaon

Chapter 11
Sharing the Future: Concluding Thoughts

Nestlé and India have a one hundred year old story to tell about how it was, it is and it will continue to be possible to Create Shared Value opportunities for the myriads of people who have taken part in this continuing journey. As the company prepared its response to the numerous challenges it needed to start the operation of the Moga factory, Nestlé engineered a corporate strategy that was at least 40 years before its time and which successfully matched corporate growth prospects with extensive community development.

Fifty years after the then unnamed business model was first put in place, Nestlé's engagement in Moga stands out as a pioneering case of how a company can position itself at the forefront of the food industry, and then successfully build and expand markets, develop quality and innovative products, and enlarge and secure value chains. All of this whilst continuously meeting societal needs and directly continuing to press forward with the social and the economic advancements of hundreds of thousands of families in, by all accounts, what was to start with a very impoverished region of Punjab. Now termed as 'Creating Shared Value,' this business practice has brought together Nestlé and the hundreds of thousands of farmers, workers, ancillary companies and local consumers that constitute the company's primary stakeholder cluster in Moga. They have built a 'community of interest' that has made the region an economically, socially and environmentally better place to live for its suppliers and consumers, and a profitable enterprise for the company.

From the moment the decision was made to open a factory in Moga in 1961, the profile of that community, its people and the neighbouring villages have changed not only profoundly but also continuously. The study has further concluded that the company has not only honoured the vision spelt out by its founder, Henri Nestlé, to positively influence the social environment and enhance quality of life, but also did so whilst and by seizing local opportunities, establishing lasting partnerships and spreading the benefits of this commitment amongst its primary stakeholder base. That is, by developing and furthering the interdependent corporate, organizational, relational and physical infrastructure, it has successfully managed to create shared value for the company, its shareholders, and the community as a whole within which it operates.

A. K. Biswas et al., *Creating Shared Value*, SpringerBriefs on Case Studies of Sustainable Development, DOI: 10.1007/978-3-319-01463-0_11, © The Author(s) 2014

The present study pieced together many of the factors, stakeholders, challenges, opportunities, solutions and foreseeable and unexpected developments that have shaped Moga since Nestlé first settled in the area. The company responded to pervasive and emerging challenges in the production, labour, price, supply and political fronts but used local conditions, potential, social capital, institutions, ingenuity and inventiveness to build, support and further Moga's comparative advantages in dairy and related industries. Nearly two decades later somewhat similar Government policies began to be formulated and implemented around the country. This study concludes that the company's growth and success has been firmly grounded on a community-centred, context-tailored, and trust-based framework without which it would not have been possible to overcome long-existing socio-cultural, religious, infrastructural and structural challenges surrounding the dairy sector and milk production and marketing at Moga.

For over 50 years, Nestlé has harnessed local endowments and agricultural potential and breakthroughs and integrated them into household, micro, regional and macro-economic and agricultural trade issues. All of these factors play complimentary and mutually reinforcing roles in long-term rural development, poverty alleviation, growing availability of improved foodstuff, raising nutrition levels, community-building and in bringing about individual, family and village employment-related benefits. It has also stepped into fill in, technical, agricultural and production skills gaps, securing both simultaneously and symbiotically, the human, physical, social, technological and political capital that were necessary to operate and continue to grow as a profitable enterprise. The result has been the entrenchment of a social value proposition into Nestlé's overall and global corporate strategy. In Moga, such efforts have greatly contributed to the profound transformation of a predominantly agriculture-based rural community that now has access to a broader and more advantageous bundle of opportunities, skills, infrastructure, aspirations and quality of life.

Nestlé induced dramatic attitudinal and cultural changes to dairying among the villagers that led to the formal organization of the milk trade and opened up direct and indirect employment opportunities for the community almost two decades before similar Government initiatives were put in place anywhere in India. By the time national policies towards self-sufficiency in food grains and the creation of a nation-wide milk grid were implemented, Nestlé and its supplying farmers already had in place a solid organizational, physical, labour and knowledge dairy infrastructure. This headstart allowed the company and Moga's community to fast track the benefits that accompanied Government efforts to boost agricultural productivity. Since then, milk has become a 'cash crop' and a source of additional and regular income for tens of thousands of landless, marginal, small and large farmers.

Without abandoning agriculture as their main economic activity, Nestlé encouraged its farmers to meet much of their day-to-day financial needs from milk. With irrigation and agrochemicals making their lands more productive, the farmers continue to sell surplus agricultural production for additional income. They also store some as cattle-feed, which in turn has increased milk yields and quality and brings in additional income from dairying. It should also be realized that as farmers have produced higher quality milk products, nutrition levels have also been on the rise as they consume foodstuffs that fulfil stringent quality standards and respond to local nutritional deficiencies. From day one, this once novel economic activity has been sustained through incremental tiers of investment, training and an extensive network of supporting services. With time, as Nestlé become more and more successful and resourceful, especially in terms of the prevailing conditions and the culture at Moga, such practices have been consolidated and extended. This trend continues and it is at the forefront of Moga's continuous and impressive transformation.

The company's experience at Moga shows business strategies can effectively remain profitable, bring about financial returns and at the same time deliver social, environmental and economic benefits to the communities associated with supporting corporate commercial activities. This study found how the company's engagement in Moga has enabled cluster development through positive spillovers and externalities, fostering endogenous and sustainable solutions to local developmental problems. With more and enhanced economic and social opportunities available to the community, residents have speed-tackled challenges related to poverty, illiteracy, seasonal unemployment, low agricultural productivity, malnutrition, lack of access to drinking water and sanitation, limited remunerated activities for women, poor infrastructure, low quality standards, etc. By the time this pioneering approach to doing business in and with a community was coined as 'Creating Shared Value', Moga already stood out as a robust and successful example of the interdependency between social welfare and corporate success.

It is seizing this success that the private sector is being urged to mobilize its resources, expertise, networks, innovative practices, research and development findings and technologies to help address the economic, social, environmental, governance, cultural, human rights and labour challenges that remain in Moga, and also the world at large. In 2000, the United Nations Global Compact launched the world's largest corporate responsibility initiative with over 10,000 private sector signatories from 140 countries. Nestlé responded to this call and voluntarily acquired the commitment to make its operations and corporate strategies support the goals the member states of the United Nations have set for the world at large, notably the Millennium Development Goals.

Moga stands as an example of how commercial objectives and social impact investment strategies can successfully, sustainably and profitably have been put in place. The more the corporate, development, non-governmental, philanthropic, academic and public sectors get to know about such successful case studies, experiences, lessons learned and initiatives, the better informed the social investment efforts these agents undertake will be in the coming years. Such case studies of sustainable development are worth carrying out. The world at large stands much to gain from these types of case studies. They can show that the problems, however entrenched and complex they may be, can be solved with a proper long-term vision, hard work, determination and an enabling environment.

Epilogue

Nestlé's sales make it the world's biggest company for fast-moving consumer goods, and its market capitalisation makes it the biggest company in the whole of Europe. When talking with people about the company, many tell us how surprised they are that a firm from a small, landlocked country could become so successful globally, in a sector that requires profound understanding of a multiplicity of local cultures and societies across the world, and comprehensive and deep integration into the local context in which it operates.

What may have looked like insurmountable challenges for Nestlé being a start-up enterprise in a small Swiss town like Vevey on Lake Geneva—especially with only a minimum of essential infrastructure available in the middle of the 19th century—have all turned out to be major assets and opportunities.

As a company from a small country without a colonial past, we could never rely on support from colonial structures. Instead, we worked from the beginning for and with the people of individual countries. When, for instance, we needed a better road to be able to transport milk quickly from farms to our factory, we had to convince the local indigenous authorities that what we do is worthwhile for the community.

Being from a small country also helped in accepting, and as a result embracing, diversity. For an American, it is sometimes strange that some people abroad do not like hamburgers; to people from Switzerland, it is quite obvious that not everybody in the world appreciates a fondue. Also here, being from a small home-country results often in advantages.

Finally, our need to locate operations close to the farmer-suppliers, outside of compounds with other foreign firms and expatriates around a country's capital, made our start more difficult at times, with certain risks for our people working there, vis-à-vis both their security as well as health. But ultimately, it facilitated and even forced our integration with the local communities and economies.

So it is the roots and history of the company, our long-term perspective, our corporate culture and ambitions, the circumstances and specificity of our sector, as also very much our people at all levels—their values, attitudes and commitment—that led us at a very early stage in Nestlé's history into Creating Shared Value, a

A. K. Biswas et al., *Creating Shared Value*, SpringerBriefs on Case Studies of Sustainable Development, DOI: 10.1007/978-3-319-01463-0, © The Author(s) 2014

concept that only got its name at the beginning of the 21st century.[1] It is based on our belief that by our way of doing business we can create at the same time value for our shareholders and for society, especially the communities in which we operate, focusing on three areas, namely water, rural development and nutrition. The long-term interest of society is built into our business case.

This book reports about such a well-embedded operation, our factory in Moga. It illustrates a model, a pattern already tested elsewhere before. When built in 1961, it was by no means the first of its kind. The Nestlé Group's first factory abroad was inaugurated in 1872 in England. It was not even the first one in developing countries—this was in Brazil in 1921.

But Moga illustrates especially well all the different aspects of the overall highly positive impact of our operations on stakeholders and communities where we invest. In particular, it illustrates that, for the region and the people, this is about a step change in livelihood, not just an incremental improvement. The new factory generated income for farmers who, before, had to quite often manage a fragile and risky state of self-sufficiency; it created direct and indirect employment in a region where there were virtually no higher value-added industrial jobs. And it provided opportunities for other entrepreneurs to start businesses, either as suppliers, or else helped by the demonstration effect.

Moga cannot be seen in isolation; it is one of 461 factories of the Nestlé Group (2011), all following in one way or another the pattern highlighted in this study. This adds up to some quite big numbers, some of which are provided below as examples. You will find more in the various editions of our Annual Report and Creating Shared Value Report.[2]

A main part of this report is about milk—and the milk district model, as the one in Moga, actually represents the most important part of our direct partnership with local farmers. Since the late 19th century, we have located our factories close to farmers in order to establish a steady and secure supply of fresh milk and economic structures. Physical collection and transportation are important, but are only part of the story. Through very reliable weekly payments to the farmer-suppliers, for instance, we helped improve the short, medium and long-term viability of the farming community. Procurement needs and issues of food safety, quality, and national priorities drove this strategy.

The main needs and strategies are still valid. Today, we have milk districts in over 30 countries across the world. In many of them, as for example in China, we are operating several milk districts.

In a number of ways, milk is special. Dairy farming is much more varied than, for instance, growing grains. We have to adapt to a huge diversity of farming systems ranging from extensive to highly intensive farming, different sizes,

---

[1] Porter, M., and M. Kramer (2006) Strategy and society: the link between competitive advantage and corporate social responsibility. Harvard Business Review 84 (12): 78–92, 163.

[2] At http://www.nestle.com/asset-library/Documents/Library/Documents/Annual_Reports/2011-Annual-Report-EN.pdf.

different technology levels and climatic zones. At Hulunbeir in China we have − 40 °C in winter, whereas in Pakistan temperatures may rise to close to 50 °C in summer.

There is also the challenge that we—and ultimately consumers—need milk of the right quality, in the right quantity, at the right time. By nature, there are seasonal variations in volumes and quality of production on the farms, and by nature milk is being produced on a daily basis. And it is also highly perishable. All this translates into high demands on infrastructure and our logistics—to ensure that the needs of our industrial structure and consumers on the one hand, and the nature of the production of milk on farms on the other, can be efficiently brought together.

In the milk districts we represent the interests of our consumers, by looking for a good and well-balanced composition of milk, ensuring safety, stimulating volume growth and keeping its cost affordable. In order to avoid these multiple demands turning essentially into pressure on the farmers, and being a process of Creating Shared Value instead, we are providing technical assistance to the farmers. The bigger part of our agricultural advisors and extension workers are in milk. In many cases, we are also providing credits and other forms of financial support.

Nestlé has, likewise, coffee factories in all the important coffee producing countries—Brazil, Vietnam, Colombia, Indonesia, China, Mexico, Thailand, Philippines, India, Ivory Coast, etc. This is a unique differentiator, as main global competitors tend to locate production facilities in countries with high coffee consumption. Whilst Nestlé has factories in these countries too, having manufacturing facilities in coffee producing countries enables us to engage directly with the farmers to source green coffee independently from trade houses. This sourcing concept is a part of "Farmer Connect"—our programme in direct procurement markets. The advantages are a shorter supply chain, less transaction costs, a direct influence on quality and a better understanding of the real issues at the farm. But it also requires a strong presence on the ground from a dedicated team of agronomists and buyers. "Farmer Connect" enables us to go beyond certification and focus on real farmers' development potentials to jointly grow the business and ensure supply to our local factories in the long run. Therefore, we have launched the Nescafé Plan, an investment of CHF 500 million in coffee projects over a period of 10 years, that involves a broad range of activities including plant breeding and distribution of 220 million high yield, disease resistant coffee plantlets to farmers by 2020, training according to the Nescafé Better Farming Practices as well as applying the fundamental sustainability criteria. Having farmers benefit from the Nescafé Plan and helping them to get higher and regular incomes will also ensure that our coffee factories will be supplied with coffee of high quality in the future, based on similar principles in the milk district model.

For another very important agricultural material, cocoa, we have launched the Nestlé Cocoa Plan. It aims to improve farm productivity and profitability and the creation of sustainable supply chains for Nestlé cocoa in countries like Indonesia, Ecuador, Mexico, Venezuela, Ghana and Ivory Coast. Between 2010 and 2019, we will invest CHF 110 million in cocoa plant science and sustainability initiatives, and distribute 12 million cocoa plantlets. In 2012 we trained over 21,000 farmers, distributed over 1 million plants, bought over 40,000 t of cocoa directly and built 12 schools through the Nestlé Cocoa Plan. Improving the economic situation of farmers will no doubt contribute to reducing child labour. And, beyond this, we engaged in 2012 with the Fair Labour Association (FLA) to report on the labour conditions in the cocoa supply chain in Ivory Coast, and to look for ways to address it effectively. Their report was published together with our action plan. One of the most important actions we committed to was to set up a monitoring and remediation process for child labour in our supply chain. Primary concrete steps for implementation were taken right away; more will follow in close cooperation with the authorities.

All in all, we partner directly with 682,000 farmers in 51 countries across the world in 'Farmer Connect': there are 423,000 milk farmers, 220,000 coffee growers and 39,000 cocoa farmers. 11,800 agricultural advisors and extension workers—part of them full-time employees, part of them on a contract basis—make sure that there is continuous improvement in the value creation by and for these farmers, in terms of production, productivity and quality of these and other raw materials needed for Nestlé products.

The milk district model, prevailing for more than 140 years, has been successfully translated, adapted and refined to other rural communities where we source directly all kind of agricultural materials. Vegetables, grains, fruits, spices, nuts, etc., are sourced in addition to the cocoa and coffee mentioned earlier. No doubt it will evolve over the next 140 years, but we are also convinced that a number of the key characteristics of this proven partnership with farmers producing milk and other food products will remain.

The report also illustrates the strong impact on employment and employee prosperity, and again, this adds up when looking at global numbers for the Nestlé Group. At the end of 2011, we directly employed 328,000 people, up from 281,000 at the end of 2010. A large number of empirical studies show that both salaries and working conditions in foreign-invested firms, particularly but not only in emerging economies, are significantly above average—and this is clearly also the case with Nestlé.[3] In addition to this direct employment with its positive impact on the families also of those employed, our operations across the world generate a large amount of indirect employment, both upstream and downstream. An overview of research shows that food processing industries generally present employment multipliers that rank among the highest in manufacturing, with the average around 4. Ultimately, individual companies have their own job multiplier defined by the coefficient of salaries versus intermediate consumption of goods and services. Due to outsourcing over the last couple of decades, this coefficient has been steadily increasing for Nestlé, probably by about 1/6 over this period. According to our

---

3 P. Buckley et al., (2009) Creating Shared Value: Economic View and Measurement; Leeds University, Leeds.

own estimates, the job multiplier may reach a value of around 4.5 for the Nestlé Group worldwide. This leads to an estimated 1.5 million direct and indirect jobs, actually measured in full-time job equivalents. That is, a much higher number of people receive part of their income thanks to linkages to Nestlé operations. And ultimately, this is not just a story about numbers; it is one about empowerment within the Group, and along the supply chain. We will illustrate this in the second part of our epilogue.

Now, we should add a remark of caution here. The company is successful and growing, and with a clear long-term strategy we are also very confident for its future. But things cannot always and everywhere only go up. Markets are changing and we have to adapt. Condensed milk is a good example here. It was, for a very long time, a highly successful product in industrialised countries too, but it has almost disappeared from the shelves in the USA and Europe now. In this case, adapting has also meant downsizing and selling production sites in order to ensure that the company maintains full strength to continue investing and expanding in other aspects of the business and in other places across the world.

As just shown, a well-embedded company creates value upstream, with and for suppliers of raw materials like the farmers, and suppliers of services and intermediate goods such as packaging materials. It creates value for employees—direct and indirect jobs resulting from a company's investment—and their families. There is one more dimension: a company well embedded in a competitive environment also creates value horizontally. One of the biggest (and intangible) assets forming the value of a company, a powerful wealth and impact multiplier, is the knowledge embedded in a company—scientific, commercial, organisational, etc.[4] This embedded knowledge, or more specifically the knowledge pool transferred by a company to a country when investing there, spreads out in these countries through so-called demonstration effects. This is about competitors watching us and often learning from us. They can be stimulated by our innovations. Supply chains (e.g., the co-operation with farmers) and products (safety, overall quality) of our competitors improve as they see how we are doing things. Quite frequently, they can also take advantage of the knowledge transferred to our suppliers, e.g., for better packaging materials. Empirical studies show substantial results here too. "Johnson and Evenson (1999) examined the effects of new technologies introduced in the agricultural and food-processing industries in 14 less developed nations and newly industrialised countries. The results of this study show that both international and inter-industry spillover effects improved agricultural productivity."[5] And, "After observing a product innovation or a novel form of organisation adapted to local conditions, local entrepreneurs recognise their feasibility, and thus strive to imitate them" (Meyer 2004: 262).[6]

---

[4] A. Toffler, Powershift, 1990.

[5] P. Buckley op. cit., p. 25.

[6] P. Buckley op. cit., p. 50.

This brings us to another major group of stakeholders—the consumers, the main winners of this horizontal value creation, is also directly in our focus. Again, here are some orders of magnitude on Nestlé initiatives from a global perspective. It all starts with knowledge and education. Our Healthy Kids programme reaches more than 6 million children and is active in more than 60 countries. The objective of the Nestlé Healthy Kids Programme, which started in 2009, is to raise nutrition, health and wellness awareness in school-age children around the world and to encourage physical activity so as to reduce the burden posed by non-communicable diseases. Another dimension is the value of our products, illustrated with micronutrients added. Some brief words on the context are important: the impact of micronutrient deficiencies can be as devastating as the one from lack of calories. According to UNICEF, between 1.9 and 2.7 million children die annually from lack of Vitamin A alone, and up to half a million a year turn blind. And the micronutrient malnutrition is a time bomb: according to Save the Children, 450 million children will fail to develop properly—both physically and mentally—due to inadequate diet. With our products, also our Popularly Positioned Products, we provide over 150 billion servings annually of micronutrient-fortified foods. This includes 35 billion servings of Vitamin A, 53 billion servings of Iron, 102 billion servings of Iodine, and 14 billion servings of Zinc annually. One serving is defined as the optimum amount of a micronutrient in the serving size as consumed, as outlined/recommended on the pack. This amount is defined in close cooperation with the health authorities of individual countries in a very focused manner, depending on the product/consumer, the region and the proven needs. Again, a word of caution: even as the biggest food company, and even with our focus on nutrition, health and wellness, our reach remains limited; the millions of other actors in global food markets have to play a role too.

Our concern for the environment starts locally, but ultimately it is truly global. We address systematically and effectively energy efficiency, the reduction of packaging materials, and steps towards more efficient transportation. Water is particularly important: there are 301 wastewater treatment plants in Nestlé factories where municipal facilities are not available or efficient enough. The very first wastewater treatment installation in the Nestlé Group was built in the early 1930s, i.e., long before the term environmental policy started to be used. At the same time, water is an issue beyond environment, and as such one of the three pillars of Creating Shared Value. The Moga study mentions our wells that provide drinking water for Indian school children in villages around the factory as one of the specific initiatives. Globally, 126 Nestlé factories are providing clean drinking water to local communities. But water requires efforts in partnership, such as the 2030 Water Resources Group, where Nestlé plays a leading role. It is a multi-stakeholder initiative to address excess freshwater use country-by-country, also looking at its larger strategic role in the economy and society. The main goal is to provide governments with a fact base, helping them to set up strategies to bring water withdrawals back into line with sustainable supply.

Though we can show some pretty significant statistics country by country as well as globally, Creating Shared Value by Nestlé is ultimately very much about the people, their lives and communities. Taking a closer look, we discover millions of individual faces of these people. Let us therefore close with a few of these stories—fill the global statistical data with life, in order to demonstrate the extent to which the realities behind our numbers and the value created are actually relevant at the level of individuals across the world; and as outlined earlier on, to show how very often our operations and products don't just lead to an incremental improvement but a step change.

One of these stories stands at the origin of the company as a tale about consumers and nutrition. Henri Nestlé, talking about his 1867 invention of infant formula, said: "When I made my discovery, I had in mind babies a few months old. But it soon became clear that the preparation was perfectly suitable for even the youngest of infants. Mrs. Wanner was seriously ill, and her child was born a month too early. He was a sickly infant, who refused not only his mother's milk, but all other types of food as well. He was convulsive, and there seemed to be little hope for him. My friend Professor Schnetzler reported the case to me and asked if he might try my food product on the infant, who was then 15 days old. Since that time, the child has been fed exclusively with my special infant cereal. He has never been ill, and he is now a "tough" seven-month-old boy, who can sit up all by himself."

A second story on consumers and nutrition comes from another branch of founding fathers of the Nestlé Group—Julius Maggi. He was concerned about the severe malnutrition of men and women workers in Switzerland in the early industrialised times. He hence developed a legume-based soup—in the form of a cube already then, though somewhat bigger than today's Maggi Cubes—as a highly nutritious, inexpensive meal, simple and quick to prepare by women who had to work long hours in factories in the middle of the 19th century.

Often it is not only about a person or a product, but about generations of families. If you visit the chocolate museum in our factory in Broc, a mountainous region of Switzerland, you will see the milk churns with the names of the families who started to supply us with milk 150 years ago, and still do—families that in the second half of the 19th century underwent a transformation like the Moga farmers did in the second half of the 20th century. And we still continue this mutually beneficial, loyal relationship with these families.

In the northernmost province of China, you find the story of a farmer who began supplying milk to Nestlé twenty years ago, i.e., shortly after we invested and built a factory in his remote region. Once he started delivering milk to the company regularly, his life changed. Originally he had only one cow. After 10 years, he owned 6 cows and 3 heifers. Thanks to a higher and more regular income, he could build a new house. And, more importantly, income from dairy farming allowed him to send his daughter and later on his son to university.

In the case of our workers, we see similar effects of this inter-generation kind. And in many countries, the sons and daughters of our employees also work at Nestlé, often in higher positions, with a better education and training that became possible due to the reliable income of the parents.

Training, combined with empowerment, is also a major driver across all hierarchical levels and functions of the company, training broadly understood, not just with an academic background. One person who can illustrate this is the recently-retired Chef of the staff restaurant at the Nestlé Mexico headquarters. More than 40 years ago, he started as a kitchen aid (*ayudante general*). Thanks to his skills as well as the continuous training provided and supported by the firm, he ultimately became Chef. In the time between starting to work and finally becoming the head of the kitchen, the number of people that had to be fed increased from an initial 300 to 1,000 today, with considerably more sophisticated menus.

Continuous training and empowerment does not stop in the kitchen, and people in our company from South Asia, Latin America and Africa as well as other regions and continents assuming responsibilities as market head and even higher positions at our headquarters illustrate this well. Here again, the stories about individuals and their achievements are more than anecdotes, and reflections of a broad reality.

Empowerment beyond the company, i.e., through the upstream and downstream employment effects, also has its own face. Our story here is of a lady in Abidjan, Ivory Coast, who has her own Nescafé shop which is styled like a big red cup and is financed by Nestlé. It is referred to locally as 'La Tasse Rouge'. The lady serves her customers at tables shaded by umbrellas, and also encourages them to enjoy her delicious homemade sandwiches that she brings to work each morning.

Empowerment of a special kind is also important in initiatives beyond day-to-day business, and beyond factory gates. Earlier on we mentioned the wells for drinking water in Indian schools. What is as important as the drinking water for the schoolchildren in a very dry and hot region is that the children themselves are in charge of the maintenance of their well!

Another real-world face of Creating Shared Value across the globe involves animal welfare and a farmer in Chiapas, one of the poorest provinces of Mexico. Some of the many measures taken to improve productivity of farmers and cows were quite simple, but effective—such as sunroofs to protect cows from the midday sun. Initially, the farmer did not like the idea: "My cows are no princesses," he said. Today he agrees that a cow feeling well gives more and better milk.

Told in isolation, these stories may sound like mere anecdotes. Studies like this book on Moga have the merit to raise individual stories to a more general level, showing the societal context of thousands of 'anecdotes' and adding them up to statistical data in and around a factory and its products. With global data across countries from our activities, the picture becomes multidimensional and is scaled up to the overall impact of the Nestlé Group. It is value created over time, looking back to the company's near-150 years of history as well as into the future, hopefully another 150 years or more. It is value along the supply chains of all our product groups. And it is value created in competitive markets horizontally. Further studies of this kind will, hopefully, confirm and deepen this knowledge about our impact.

Looking into this impact on the societies we are embedded in, reporting and generalising facts and functionalities as analysed in this study may serve several purposes.

First, it will help in understanding. As the company and societies are undergoing continuous change, we have to re-evaluate our impact from time to time in order to make sure we still fully understand its nature and scope.

Second, it will help to make sure our impact is relevant, check whether our business activities and major initiatives beyond day-to-day business actually do create real and relevant value for individuals, and additionally, see whether this value is adding up to something that truly matters for the communities of stakeholders around our operations in individual regions and countries. We have to always bear in mind that, across the world, we are only one among millions of companies in and beyond our sector generating prosperity.

Third, it will help to communicate. Measuring, evaluating and describing Creating Shared Value may bring to a wider public some of the understanding of what industry is about.

Finally, it will ensure continuous improvement so the value created becomes more significant, both in quantity and quality. Measuring and evaluating the broad outcomes of our business activity and our actions beyond, in a broader context, can help with the efficient allocation, and where necessary re-allocation of resources and, in particular, with overcoming piecemeal approaches. It can guide our long-term efforts to serve both shareholders and society by continuously creating even more value.

Peter Brabeck-Letmathe, Chairman, and Paul Bulcke, CEO
Nestlé SA

Nestlé Reports

Food Specialities Limited (1965) Report & Accounts for the Year ended 31st December, 1964, Directors' and Auditors' Report, New Delhi
Food Specialities Limited (1966) Report and Accounts for the Year ended 31st December, 1965, Directors' and Auditors' Report, New Delhi
Food Specialities Limited (1967) Report and Accounts for the Year ended 31st December, 1966, Directors' and Auditors' Report, New Delhi
Food Specialities Limited (1968) Report and Accounts for the Year ended 31st December, 1967, Directors' and Auditors' Report, New Delhi
Food Specialities Limited (1969) Report and Accounts for the Year ended 31st December, 1968, Directors' and Auditors' Report, New Delhi
Food Specialities Limited (1970) Report and Accounts for the Year ended 31st December, 1969, Directors' and Auditors' Report, New Delhi
Food Specialities Limited (1971) Report and Accounts for the Year ended 31st December, 1970, Directors' and Auditors' Report, New Delhi
Food Specialities Limited (1972) Report and Accounts for the Year ended 31st December, 1971, Directors' and Auditors' Report, New Delhi
Food Specialities Limited (1973) Report and Accounts for the Year ended 31st December, 1972, Directors' and Auditors' Report, New Delhi
Food Specialities Limited (1974) Report and Accounts for the Year ended 31st December, 1973, Directors' and Auditors' Report, New Delhi
Food Specialities Limited (1975) Report and Accounts for the Year ended 31st December, 1974, Directors' and Auditors' Report, New Delhi
Food Specialities Limited (1976) Report and Accounts for the Year ended 31st December, 1975, Directors' and Auditors' Report, New Delhi
Food Specialities Limited (1977) Report and Accounts for the Year ended 31st December, 1976, Directors' and Auditors' Report, New Delhi
Food Specialities Limited (1978) Annual Report, 1977, Directors' and Auditors' Report, New Delhi
Food Specialities Limited (1979) Annual Report, 1978, Directors' and Auditors' Report, New Delhi
Food Specialities Limited (1980) Annual Report, 1979, Directors' and Auditors' Report, New Delhi
Food Specialities Limited (1981) Annual Report, 1980, Directors' and Auditors' Report, New Delhi

A. K. Biswas et al., *Creating Shared Value*, SpringerBriefs on Case Studies of Sustainable Development, DOI: 10.1007/978-3-319-01463-0, © The Author(s) 2014

Food Specialities Limited (1983) Annual Report, 1982, Directors' and Auditors' Report, New Delhi
Food Specialities Limited (1984) Annual Report, 1983, Directors' and Auditors' Report, New Delhi
Food Specialities Limited (1985) 26th Annual Report, 1984, Directors' and Auditors' Report, New Delhi
Food Specialities Limited (1986) 27th Annual Report, 1985, Directors' and Auditors' Report, New Delhi
Food Specialities Limited (1987) 28th Annual Report, 1986, Directors' and Auditors' Report, New Delhi
Food Specialities Limited (1988) 29th Annual Report, 1987, Directors' and Auditors' Report, New Delhi
Food Specialities Limited (1989) 30th Annual Report, 1988, Directors' and Auditors' Report, New Delhi
Nestlé India Limited (1990) 31st Annual Report, 1989, Directors' and Auditors' Report, New Delhi
Nestlé India Limited (1991) 32th Annual Report, 1990, Directors' and Auditors' Report, New Delhi
Nestlé India Limited (1992) 33rd Annual Report, 1991, Directors' and Auditors' Report, New Delhi
Nestlé India Limited (1994) 35th Annual Report, 1993, Directors' and Auditors' Report, New Delhi
Nestlé India Limited (1995) 36th Annual Report, 1994, Directors' and Auditors' Report, New Delhi
Nestlé India Limited (1996) 37th Annual Report, 1995, Directors' and Auditors' Report, New Delhi
Nestlé India Limited (1997) 38th Annual Report, 1996, Directors' and Auditors' Report, New Delhi
Nestlé India Limited (1998) 39th Annual Report, 1997, Directors' and Auditors' Report, New Delhi
Nestlé India Limited (1999) 40th Annual Report, 1998, Directors' and Auditors' Report, New Delhi
Nestlé India Limited (2000) 41st Annual Report, 1999, Directors' and Auditors' Report, New Delhi
Nestlé India Limited (2001) Annual Report, 2000, Directors' and Auditors' Report, New Delhi
Nestlé India Limited (2002) Annual Report, 2001, Directors' and Auditors' Report, Gurgaon
Nestlé India Limited (2003) Annual Report, 2002, Directors' and Auditors' Report, Gurgaon
Nestlé India Limited (2004) Annual Report, 2003, Directors' and Auditors' Report, Gurgaon
Nestlé India Limited (2005) Annual Report, 2004, Directors' and Auditors' Report, Gurgaon

Nestlé India Limited (2006) Annual Report, 2005, Directors' and Auditors' Report, Gurgaon

Nestlé India Limited (2007) Annual Report, 2006, Directors' and Auditors' Report, Gurgaon

Nestlé India Limited (2008) Annual Report, 2007, Directors' and Auditors' Report, Gurgaon

Nestlé India Limited (2009) Annual Report, 2008, Directors' and Auditors' Report, Gurgaon

Nestlé India Limited (2010) Annual Report, 2009, Directors' and Auditors' Report, Gurgaon

Nestlé India Limited (2011) Annual Report, 2010, Directors' and Auditors' Report, Gurgaon

Annex 1: Nestlé in Moga: Timelines for progressive development, 1960–2012

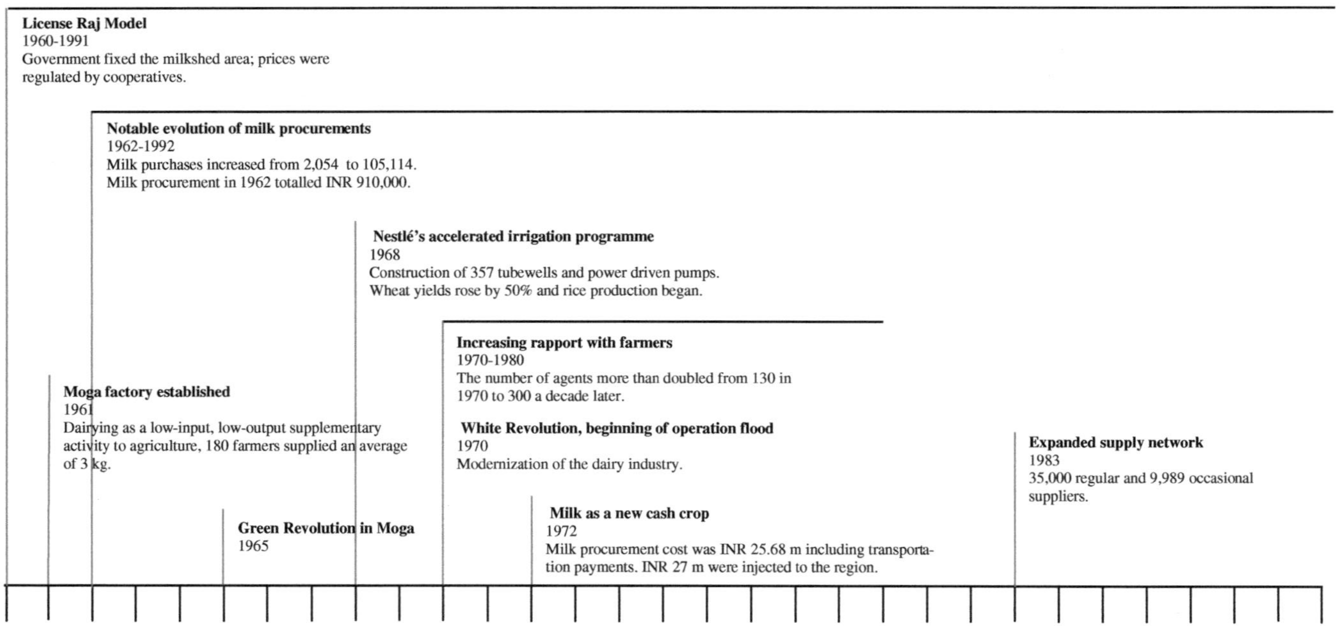

A. K. Biswas et al., *Creating Shared Value*, SpringerBriefs on Case Studies of Sustainable Development, DOI: 10.1007/978-3-319-01463-0, © The Author(s) 2014

Liberalization of the Indian economy
1991
Abolishment of Government fixed prices and milksheds.

Contributions to the local economy
1992
Nestlé was paying INR 609 m a year in milk procurements.
This was only INR 910,000 in 1962 and slightly over INR 25 m in 1972.

Key economic actor
1992
Injected INR 609 m for milk payments and INR 80 m in local taxes, commissions to milk agents and in transportations costs.

Environmental Sustainability
1997-2012
Reduction of 73% in water use, 78% in wastewater generation, 65% in energy use, and 70% in CO_2 emissions per tonne of production for Nestlé India.

Community activities
1999-2012
Water facilities have been constructed in 116 schools, benefiting over 40,000 students. Water awareness programmes have reached 20,565 students.

Intensified competition for milk procurements
2002
11 state cooperatives, 30 private dairy plants.

An important tax payer
2009-2010
INR 507.7 m in taxes were paid to the state Government and INR 163 m to the central Government.

Furthering socio-economic development
2010-2011
Nestlé payments to milk producers alone exceeded INR 6,120 m each year.

Extended network of ancillary services
2011
Nestlé's 147 main suppliers employed 86,200 persons.

Moga's consolidated and strong dairy tradition
2012
Punjab contributes to 10.5% of India's milk production with 3% of the dairy population.
Milk procurements reached 355,000 t.

Annex 2: An Evolving Landscape: Nestlé in Moga, 1961–2012

Local context	Early years, 1960–1970	1980s	1990s	2011 and beyond
National development plans or strategies	License Raj model: Government planned economy and regulated production. Self-sufficiency oriented, import substitution industrialization model 1966—The Green Revolution and progress towards self-sufficiency in food grains; introduction of high-yielding varieties of seeds; increased use of agrochemicals and irrigation 1970—Operation Flood, the world longest running development project, was put in place to create a nationwide milk grid	Self-sustaining White Revolution—Operation Flood to modernize the dairy industry. Introduction of Government-sponsored farmers' cooperatives	Trade liberalization and economy de-regularization Development of a competitive agricultural sector: de-licensing and de-canalization of dairy products	Following strong economic reforms and liberalization (opening to international trade and investment, deregulation, tax reforms and inflation-controlling measures) the country continues to witness increases in the incomes of people. In spite of economic and political turmoil, the current rate of growth is significantly above average by international standards
Risks and uncertainties	In 1966 and 1967, currency devaluations; droughts; and land and labour problems led to an increase on production costs and a surge in imported raw materials Labour setbacks were reversed due to continued efforts to provide extension services to farmers to increase yields and quality	In Moga and neighbouring districts, from 1986 to 1993, 200 deaths and over 25 people injured were attributed to terrorist activities. Throughout this period, the good rapport Nestlé had established with the local civil administration and the key economic role it played, and still does, for many local families, meant that the number of milk suppliers continued to grow in all but one year. In 1988, milk supplies decreased as terrorism escalated, restricting the inflow of labourers from Bihar and Uttar Pradesh and triggering the outflow of local residents and migrant workers That same year, floods affected 200,000 people in over 1,483 stranded villages		
Local economic landscape and living standards	In 1960, Punjab was a place of abject poverty; widespread malnutrition; high population growth; mud houses; poor transportation; very few houses had access to electricity; low productivity; and subsistence agriculture as main economic activity. Water supply, irrigation systems and transportation were mostly animal-operated With the Moga factory opening in 1961, soon a milk economy developed, the dairy sector was organized, and ancillary economic services started to mushroom	Nestlé has become a key player in integrating household, micro and macro-economic and trade issues and agricultural potential influencing poverty alleviation and reduction efforts, nutrition, community building and individual, family and community employment-related benefits. With extra income receipts, some large farmers have expanded and furthered their agro-business activities, investing more in better farming and animal husbandry inputs (feed, vaccinations and medicines) as well as in other economic activities The expanding range of goods produced in the Moga factory has brought about a similar increase in local employment opportunities due to the expansion of supply networks and market value chains for raw and processing materials, as well as for related services. This suggests employment opportunities have not only been created for milk suppliers and farmers, but also for factory employees, contractual and daily labourers and agents. Moreover, the quasi-continuous factory extension since 1980 has provided around 350 jobs annually for Indian construction companies Hoteliers, builders and various other craftsman and entrepreneurs have now settled in Moga to seize the business opportunities that have been created following the successful development of the area		
	In 1962 Nestlé contributed INR 952,000 in milk payments, increasing to INR 29 million in 1972. By 1974, Nestlé was purchasing CHF 22 million worth of local raw materials (Nestlé 1978)[a]	By 1980, 210,000 local persons were associated with Nestlé, benefitting in one way or another from the company's various development programmes In 1983, milk deliveries came from 35,000 regular and 9,989 occasional suppliers. By 1987, this number increased to 84,585 farmers	In 1992, Nestlé contributed INR 740 million to the local economy in milk payments to its farmers. Most of this disposable income was spent locally. Moreover, INR 80 million were paid in local taxes, commissions to dairy milk agents, and in transportation costs In 1992, around 77 workshops were delivering electric and mechanical support to the Moga factory	In 2011, the commercial transactions between Nestlé and 147 related enterprises were responsible for the formal employment of over 86,200 people.
	Most regular suppliers continue to be small and very small landholders and as such, the increased disposable income they receive from milk sales represented a significant complement to their main farming activities. Milk payments bridge the income gap from selling wheat in April and rice in October and as farmers usually obtain their income from selling their crops twice a year, they have started to rely on regular and predictable fortnightly milk payments from Nestlé for daily expenses Marginal milk suppliers constituted a significant group. Since they are economically disadvantaged, milk receipts are of great benefit to them as there are no caps on the amount of milk supplied. They rely on Nestlé as a guaranteed buyer for their produce. More generally, albeit depending on the season (and thus fluctuating farm earnings), milk-related income is used for feed and fodders; in improved farm equipment and agricultural inputs, for household foodstuff consumption and clothing; short-term savings and repayment of debts and loans; school fees; socio-cultural activities (weddings, festivals); demerit goods (alcohol and tobacco); and improved family and animal health			

(continued)

A. K. Biswas et al., *Creating Shared Value*, SpringerBriefs on Case Studies of Sustainable Development, DOI: 10.1007/978-3-319-01463-0, © The Author(s) 2014

(continued)

Local context	Early years, 1960–1970	1980s	1990s	2011 and beyond
Local dairying conditions	The sector was largely unregulated and subject to adulteration and arbitrary practices. Dairying was a low-input, low-output, low-income supplementary activity to agriculture Government established floor prices for milk processing factories Nestlé's engagement in Moga revolutionized socio-cultural attitudes to dairying including the removal of stigma attached to the commercial production and sale of milk and the establishment of a context-tailored milk collection system In 1961, the company's supply chain initially drew inputs from 180 farmers in four villages. By 1970, this network had reached out to 700 villages, each one engaging an average of 250 families	The company cooperatives' rates as floor prices and paid a premium 9–12 % higher. Additional bonuses between 7 and 10 % were introduced to encourage farmers to feed their animals better during low season, thereby reducing supply fluctuations Even then, Nestlé was buying less than 25 % of Moga's rising total milk production; which from 1982 to 1992, rose from 265,000 to 467,000 t[b]	Following national economic reforms and liberalization, the Milk and Milk Products Order (MMPO) came out in 1992. It came to regulate the local production of milk and dairy and included sanitary and hygienic regulations to ensure product quality Throughout the decade, Nestlé's milk share remained at around 22.5 % even when new stakeholders entered the local market. By 1991, the Government cooperatives Milkfed started setting up plants in Moga, buying 5.9 % of the district's total milk produced by 1992. Other private companies settled in the area, taking 3.1 % of all yields	In 2012, Punjab now has about 3 % of Indian dairy population but contributes to 10.5 % of the national dairy production[c] By 2002–2003, 11 Government cooperatives, 30 private dairy plants and a few public-private ones were operating in Moga. Nestlé aside, these new stakeholders were buying 5,480 t of milk[d]
	Nestlé consistently set higher quality standards than those domestically imposed for both its inputs and final products. It has also provided information on how to meet, maintain and improve such higher standards. These quality upgrades have been used as quality benchmarks and common denominators for other stakeholders involved in dairy production			
Resource endowments and infrastructure	Amelioration of poor, weather-dependent infrastructure, absent or severely limited electricity supply, poor irrigation systems and animal-operated water lifting devices In 1968, Nestlé began an accelerated irrigation programme consisting of drilling 357 tube wells and installing power driven pumps, which contributed to increases in wheat yields by 50 %. Rice could be grown from July to October This and similar actions prepared the area to quickly reap the benefits of the Green Revolution and mechanized and modernized agriculture	National and state-level macroeconomic growth has led to an overall improvement in the communities' physical capital and available infrastructure. This process furthered consolidated the Moga factory's beneficial impact in the region as a permanent source of local wealth creation: All villages have been connected to the electricity grid Increasing use of landline and mobile phones Growing number of paved roads Easier accessibility by land Modernization of the transportation system Steady increase in the number of motorized vehicles Mechanization of irrigation systems and fodder-chopping activities Establishment of milk parlours		

(continued)

(continued)

Local context	Early years, 1960–1970	1980s	1990s	2011 and beyond
Human and community development context	In 1961, only a handful of villages had primary schools and most of the population was illiterate. Soon after the Moga factory began operating an elementary school was established in Dudhike, and was later upgraded to high school in 1966–1967. By 1968, the 1st graduate college opened in Moga and many female students enrolled. There was a surge of English secondary schools	The company carried out periodical studies assessing the progress of the dairy industry in the region and the socio-economic challenges faced by the communities and that may impact milk production. Such evaluations have indicated the multiplier effects the company has had in the transformation of the region and on the overall improvement of the communities' living conditions:		
		Improved living conditions with brick houses and cement flooring		
		Construction of dry-pit latrines and indoors sanitation facilities		
		Enhanced nutrition derived from higher quality in the supply of raw materials, that are also used for household consumption, as well as from the growing availability of quality finished products		
		Higher literacy rates, growing demand for more schooling facilities and better education. Still, in 2011 overall literacy rates for Moga (71.6 %) were lower than those at the provincial level (75 %). However, the local literacy rate for women was 67.4 % compared to the Punjab average of 63.4 %[e]		
		Household adoption of more hygienic food and nutrition practices and habits		
		Establishment of an entrepreneurial class		
		In close collaboration with India's National Dairy Research Institute and Punjab Agricultural University, awareness campaigns have led to the implementation of good practices in animal husbandry, pesticide residues reduction, fuel efficiency and use of biogas plants		
		Intensified use of biogas plants and smokeless stoves, replacing wood or coal ones. By 1987, Nestlé had already constructed 2,285 smokeless stoves in the region		
		By 1992, one in 10 villages had secondary schools and one in five major villages had high schools		During the last 10 years, more than 15 colleges of engineering, pharmacy, etc. have been opened in and around Moga
Socio-cultural and gender-related attitudes	Taboos regarding milk commercialization discouraged sustaining or rising production levels. Overtime, this belief was abandoned and milk production began to be accepted as an income-generating activity	By working with the local community to gradually build trust and confidence in the milk trade, Nestlé's factory in Moga kick-started what is now a 'strong dairy tradition.'		
	Nestlé's veterinary specialists, agronomists and agro-technicians greatly helped in changing fatalistic attitudes towards illness and diseases that afflicted cattle	Farming and dairying activities seem to be sharply divided along gender lines. Women take care of cutting and chaffing fodder; animal feeding, watering, cleaning and care; and milking. A reduced number is involved in delivering milk to collection centres. As they carry out the bulk of dairying work, women's training needs has been addressed through the 'Village Women Dairy Development Programme.' These schemes have benefitted over 51,000 women and aim at improving milk quality and yields, animal hygiene. As literacy rates remain comparatively low among women, albeit higher than the provincial average, capacity building relies heavily on audio-visual media for all educational purposes: diagrams, charts, illustrations, posters, slides, photos, pictorial-guides, pre-recorded cassettes with lyrics, and short films. The transmitted skills seek to encourage them to get further involved in milk production and other remunerated activities		
	Growing recognition of the roles women play in the dairy industry. They were responsible for over 87 % of milking and animal watering, cleaning and feeding[f]	Women have been targeted by nutrition campaigns for households to adopt more balanced diets, and better hygiene and sanitation practices. Some community members have reshaped gender expectations by taking on professional roles in male-dominated activities such as tractor driving		
		However, despite encouraging changes in gender labour divisions and women's rising involvement in milk production-related activities, there is still a long way to go to further their participation in the formal economy and income generating activities available around Moga		

(continued)

(continued)

Local context	Early years, 1960–1970	1980s	1990s	2011 and beyond
Nestlé's rapport with farmers	The company used cooperatives' rates as floor prices and paid a premium 9–12 % higher. Additional bonuses between 7 and 10 % were introduced to encourage farmers to feed their animals better during low season. This reduced supply fluctuations The company, supplying farmers, factory workers and ancillary services established long-term social capital and loyalty-based, relational transactions		Market changes and increased competition have resulted in farmers' rising milk supply elasticity. Profitability has come to be based on the price of milk alone. This is particularly true among young farmers, whose attitudes and expectations regarding the extent to which corporate success ought to be tied to the community's wellbeing have been fully internalized as part of normal business activities. Conversely, elder farmers point at the 50-year business relationship with Nestlé and seem to have more deeply internalized the transactional and relational benefits resulting from this partnership	
	Initially driven by the need to increase milk intakes, Nestlé actively sought to help the farmers in increasing quantity and quality of milk. Overtime, the company and farmers supplying it have established a 'community of interest' in which the company encourages, supports and facilitates those factors necessary for its operation which in turn are also factors spiralling Moga's development process:			
	Highlighting relational over transactional interactions. Nestlé Agricultural Extension Services keeps direct and close contact and rapport with over 85,000 individual farmers as experts and trainers spend most of their time in the fields			
	Relying on the establishment and strengthening of mutual and effective communication avenues between the company and supplying farmers, farmers and veterinarians, farmers and agronomists and farmers and extension departments			
	Offering systematic and/or specialized factory training, apprenticeships and postings in other factories abroad			
	Encouraging local banks to extent credit lines for dairying related activities and inputs			
	Putting in place affiliation incentives for milk suppliers as payments for required inputs and deducting loan repayments from milk payments			
	Equipping farmers with transferable skills that can help them perform different and better tasks, enlarging their occupational flexibility and mobility and their engagement in additional income-generating activities			
	Publicly recognizing successful farmers, disseminating their achievements and encouraging other entrepreneurs to follow			
	Drawing in local talent, recruiting workers from the community and offering extensive training, and			
	Using local knowledge and partnering with local institutions, universities and banks to foster and disseminate innovations			
	Factory employees, agents and dairy producers working with and for Nestlé have benefitted from the reputational capital and social prestige associated with the brand. As ancillary firms adopt quality standards and industrial processes meeting Nestlé's requirements, they can also trade with other companies who see their association with Nestlé as a quality and reliability guarantee. Working and collaborating with the company has had an aspirational impact in the community			
	Continuous and expanding network of free, relevant and cutting-edge veterinary, husbandry, dairying and agricultural advice; technical assistance, education and capacity building sessions; and supporting services that:			
	Are delivered by context, region and local conditions-sensitive specialists			
	Support research, innovation, development and dissemination of successful practices regarding key inputs and raw materials			
	Encourage the adoption of novel and innovative farming methods			
	Promote peer-to-peer learning among suppliers			
	Foster better animal health and preventive treatments			
	Help close farmers' skill gaps affecting the supply of raw materials to Nestlé			
	Extensively involve Nestlé staff at suppliers' locations to ensure successful implementation of new technologies and quality control processes			
	The company offers its factory employees a competitive package that goes well beyond Government requirements of 48 working hours per week, 14 days of holidays and medical insurance and treatment for each employee. Nestlé's workers receive four times as high salaries compared to the Government-fixed minimum salary for 45 working hour weekly shifts; seniority-based progressive holidays; medical insurance and treatment for employee and their families; and canteens services at nominal prices. Absenteeism is low (4.5 %) and so is the turnover rate, 1.6 %			
	Farmers have favoured supplies to Nestlé due to: the prices paid for milk and other raw products; steady and timely payments that act as insurance against price fluctuations and smoothen income receipts; convenient payment methods; accurate, strict and fair quality assurance controls and measurements; supporting services and training; offer of financial services; trustworthiness and responsiveness of accountability mechanisms; corporate reliability, trustworthiness and reputability; and expectations of positive technological, managerial, technical and skill spillovers			

(continued)

(continued)

Local context	Early years, 1960–1970	1980s	1990s	2011 and beyond
	Some ancillary enterprises have developed a symbiotic relationship with Nestlé, tying most of its success and business expansion to that of its business partner. Paras Spices PVT Ltd. has built a systematic collaboration with Nestlé, establishing a wide and solid range of interactions and mutually beneficial investment and training requirements. The scope of their transactions has grown from a 2 kg turmeric purchase in 1982 to one involving over 5,000 farmers directly involved in the production of spices from all over the country and 35,000 to 40,000 more indirectly involved in the process. This same relation has brought Para's employees a competitive income, training, medical plan, safety and security package. The spices company expansion ventures are supported with Nestlé's know-how, experience in factory design and laboratory requirements, and technical assistance that can be easily reached due to geographically proximity between the two partners			
	Transportation and animal-feed companies have also spurred and expanded their business operations by harnessing the market opportunities being associated to Nestlé has opened for them. Brar Enterprises Limited has been transporting milk to Nestlé since the factory was first established and since then, the company has used its with in-house capital to upgrade, improve and adapt its equipment and technology to respond to Nestlé's requirements, expectations and standards. This raised Brar Enterprises' business practices quality, setting a much higher floor thereby made available to local and multinational companies alike making exceptionally good quality the transporters' status quo practice. Nestlé has gone from being Brar's sole partner to the current 20 % of business activities			
	Before the arrival of Nestlé in Moga, cattle-feed as an organized business activity was unknown and so fodder was sold to farmers on a non-profit basis. However, as quantity and quality demands for animal feed rose, local entrepreneurs stepped in. P. Marka Registered had been producing mustard oil since 1933 and as it also sold mustard cakes, it decided to enter the cattle-feed business in 1988. Initially, their plant original capacity of 50 TPD has now been doubled. The company now employs over 25 trained people working entirely on preparing animal feed according to a Nestlé-provided formula and meeting the required quality standards			
	The company has been fully integrated in the social, cultural and economic life of Moga, quickly responding to immediate and long-term needs that strengthen the links between Nestlé and the community in which it operates. For instance, during the 1988 floods that afflicted the region, Nestlé was at the forefront of relief operations, distributing food, fodder, and veterinary services to the stricken people. Since 1999, longer-term and sustainable community development programmes have been undertaken to promote prosperity in the community. Around 115 of these initiatives have provided drinking water and other basic facilities at local schools, benefitting over 40,800 students. Other programmes have supported immunization campaigns and set medical camps. Whilst the company finances such projects, it encourages community ownership and local responsibility for maintaining them			

[a] Nestlé in the Developing Countries, 1978, Nestlé Alimentana S.A., Vevey, p. 139

[b] Nestlé in India, 1962–1992, pp. 33–34

[c] Interview with Dr. Surinder Singh, Director Research, Guru Angad Dev Veterinary and Animal Science University (GADVASU), Ludhiana on 14/11/2011

[d] Dairy Scenario of Punjab, Gill, M.S., Bhatti, J.S., available at http://pdfa.org.in/website/english_articles/DAIRY%20INDUSTRIES%20IN%20PUNJAB.pdf (Accessed on 29/5/2012)

[e] Parvesh, Sharma. "Punjab females less educated than males: SSA." Times of India, November 11th, 2011. Web. June 7th, 2012. http://articles.timesofindia.indiatimes.com/2011-11-07/india/30369061_1_literacy-rate-males-punjab-education-department

[f] Education and information on Animal Husbandry Techniques for Rural Women: Why and How, T. S. Sandhu, FSL, Moga (no date), pp. 4–5

Annex 3: Evolution of Developments, 1961, 1981 and 2011

		1961	1981	2011	2012
	Corporate profile				
1.	Moga factory processing capacity	The processing capacity has gone up by about 18 times between 1961 and 2011			
2.	Goods produced	Condensed milk		Milk products and drinks (baby formula, ghee, yogurt); prepared dishes and cooking aids (noodles, soups, spices, sauces); and chocolates and confectionery	Milk products and drinks (baby formula, ghee, yogurt); prepared dishes and cooking aids (noodles, soups, spices, sauces); and chocolates and confectionery
3.	Total production for Moga factory	The total production of the factory has gone up by 95 times between 1961 and 2011			
4.	Water use per tonne of final products			9.70 (m^3/t)	9.5 (m^3/t)
5.	Wastewater generated per tonne of final products			8.57 (m^3/t)	9.3 (m^3/t)
6.	Energy consumed per tonne of final products			7.88 (GJ/t)	8.2 (GJ/t)
	Milk distribution and procurements				
7.	Milk routes	3	13	31	31
8.	Collection centres (in Punjab only)	63	434	1,568	1,483
	Milk supplies				
9.	Calf mortality rate	70 %	50 %	25–30 % (natural 5 %)	20 % (natural 5 %)
10.	Direct suppliers	180	32,000	68,846	62,183
11.	Total volume supplied in kilos	511	35,000,000	211 m kg	236 m kg
12.	Average farmer milk supply	2.83	1,093.75	3,065 kg	3,714 kg
13.	Milk productivity				
	Buffaloes	5.3 kg/day	6.5 kg/day	6.0 kg/day	8.0 kg/day
	Cows	3.0 kg/day	8.0 kg/day	15 kg/day	15 kg/day
14.	Average milk price in INR per kg of fat	0.45	34	363.51	393.46
	Farmers' profile, % of total suppliers				

(continued)

(continued)

		1961	1981	2011	2012
		1961	1980–1981	2009–2010	
15.	Farm labourers		27.31	18.5	Study on farmer profile was not carried out in 2011–2012. It is carried out after every 3–4 years
16.	Marginal		40.5	27.47	
17.	Small		22.56	24.94	
18.	Average		22.56	21.35	
19.	Large		0.95	5.36	
20.	Very large		0.2	2.39	
		1961	1981	2011	
	Employment generation				
21.	Direct jobs (factory)	292 (for the year 1968)	589	1,192	1,189
22.	Indirect jobs in ancillary industries	—	—	86,200	
	Contributions to the local economy				
23.	Milk payments	INR 0.91 m (in 1962)	INR 75.75 m	INR 6,122 m	INR 7,401 m
24.	Payment to milk collectors	—	INR 5.1 m (in 1982)	INR 147 m	INR 172 m
25.	Payments to factory employees	—	—	INR 657 m	INR 770 m
26.	Taxes paid to the state of Punjab	—	—	INR 553 m	INR 540 m

Annex 4: How Nestlé has Touched the Lives of People

Case Study 1: Jatinder Singh, Gureh Village

Jatinder Singh, like most other youths of his village Gureh, was a hard-working farmer. He inherited one buffalo from his father. Life was difficult for a very small farmer like him. It was his hard work and the support of his mother and wife that assured the survival of his 5-person family.

Stories of NRI (Non-Resident Indian) villages such as Dudhike are well-known in the Moga area. People had started sending their children to English medium schools for better education so that they can go to cities, or even abroad, to earn a better livelihood. Jatinder Singh was no different but with his meager income it simply was not possible for him to send his children to good schools.

By 1990, with his hard work and frugal living, he managed to purchase three more buffalos and started selling milk after being encouraged by one of Nestlé's field staff. This field staff convinced him that he could use this additional, continuous and reliable income for his children's education and meet his day-to-day expenses.

Before he could start selling his milk to Nestlé, a local agent of Verka (Punjab State Cooperative Milk Producers Federation), a governmental institution poached him by offering higher price for his milk and a hassle-free acceptance of his milk. He assured Jatinder Singh that he would not bother him about the quality of milk.

Soon Jatinder Singh becomes disillusioned with the agent. This agent at village Karnail Singh was often drunk. After first 2–3 regular payments, the agent started to delay in paying him on one pretext or another. Soon he started paying less than what was promised. Not only this, but the agent also started accusing Jatinder for supplying adulterated milk.

A dejected Jatinder thought about not selling milk to anyone any more. Then, his mother heard that Nestlé was organizing a factory tour for the women of villages. She was curious to see the factory. When she visited the Moga factory and interacted with staff and officers and saw herself the plant and its management, she returned home very impressed. Apart from cleanliness, transparency and professionalism, what attracted her most was the attitude of the staff, who sat down with her and discussed her problems. She was amazed to see how easily and respectfully Nestlé staff treated her and gave her helpful suggestions about nutrition of humans and animals and other dairy-related issues.

A. K. Biswas et al., *Creating Shared Value*, SpringerBriefs on Case Studies of Sustainable Development, DOI: 10.1007/978-3-319-01463-0, © The Author(s) 2014

This changed her perception about Nestlé. In 1999, the family began to supply milk to the company. Later, Nestlé's staff helped them with technical advice to buy high yielding cows. With Nestlé's technical assistance and regular guidance by its staff, Jatinder's cattle stock started to increase. Observing his sincerity and hard work, Nestlé further encouraged him to take charge of an agency. By March 2005, his agency was supplying 3,500 kg of milk to Nestlé.

Nestlé then sent Jatinder's son along with others to a Dairy Expo in the USA, where they saw mechanization of farms and interacted with dairy farmers from all over the world.

After returning from the USA, and again with active guidance from Nestlé, Jatinder and his son installed a milking parlour, constructed better animal-sheds and began scientific management of calves. With the growing size of farm, Jatinder also focused on quality silage-making. Within ten years, he had 85 animals out of which 50 were adult milking cows.

Expansion of his own farm forced him to leave the milk collection centre so that he could focus better on milk production. Today, his farm yields daily 500 kg of milk which he hopes to double in the future.

He said categorically that without the help of Nestlé, he would have remained a small poor farmer. His son is now working in Canada and his daughter is pursuing a diploma course in Mechanical Engineering. (Interview with Jatinder Singh at his farm in Gureh village, 18th November, 2011)

Case study 2: Jagdeep Singh, Assal village

The case of Jagdeep Singh, Assal village, is one of many that the team interviewed whose life has been completely transformed by Nestlé's activities.

Punjab is well-known for its hard-working peasants and gallant soldiers. The tradition of serving in the armed forces and farming has been persistent for centuries in rural areas. Generally, one son remained at home farming and the other joined the armed forces. This tradition led to the slogan of Jai Jawan Jai Kisan (*Victory to the soldiers, victory to the farmers*), coined by India's second Prime Minister, Lal Bahadur Shastri.

When Nestlé opened the Moga factory in the early sixties, it was not a good time for India. The country was dependent on food imports. The moral was low because of its poor performance in the war with China in 1962. Food shortages had forced the Prime Minister to call for fasting. Industrial developments depended on an inefficient public sector. Private sector was heavily controlled by red tape.

When Nestlé opened its factory in Moga, it was not the best time for the region. With the beginning of the Green Revolution in 1965, and tepid industrial development, socio-economic character of the villages began to change fairly rapidly. These new forces required more working hands at home. Accordingly, young males stayed at home for farming which became economically attractive.

As employment opportunities started to increase, parents became reluctant to send their sons to the army. Marginal farmers and labourers who had always encouraged their sons to join army suddenly had better local options.

Jagdeep Singh's father was in the army, but he did not allow him to join the army. According to Jagdeep, his father said, "I had always been away from home during my service. But those were different times. There was nothing for poor people like us to earn a living in the villages. Now, the situation has changed. You can get better jobs near home". Thus, he joined the police in 1998. As soon as he had a job, his parent arranged his marriage in 1999, when he was 23. With new responsibilities, his expenses started to rise and he found the salary inadequate to meet his needs. He informed us:

"My basic pay was INR 1,827 and I used to get a total sum of INR 4,500. With this meager amount, I had to run the house, take care of my parents and wife. I used to worry how much should I give to my parents and how much should I keep for myself? How can I rear my children properly? Many times, I thought of quitting my job and going back to the village. But my confidence level and courage were low. Some time ago my father had purchased some rural land using his retirement funds but I had no idea how I could improve my economic situation."

Leaving a salaried and secure job was a difficult decision. He shared his problem with his Inspector. One day, during a visit to Punjab Agricultural University, he met Dr. Jagtar Singh, head of its Horticulture Department and a relative of his Inspector.

When Dr. Jagtar Singh heard that he had 7 ha of land he told Jagdeep that if he had that much land, he would have left as the Head of Horticulture Department. He advised Jagdeep, "you are young and can do hard work. When you grow old and retire without taking any risk, you will only regret. Take the risk, resign and start on your own. You would be a success in 10 years".

Jagdeep resigned from his job. He recollects:

"His words inspired me. Today, I realize how prophetic his words were. Currently my batch-mates in the police are getting a salary of INR 18,000 per month, I now have 13 employees who receive a salary of INR 5,000 or more per month. I now have 35.5 ha of agricultural land where I grow potatoes, barseem, wheat, chilli, garlic and peas. This and a big dairy farm producing 11 quintals of milk per day: all in a span of 10 years".

Dr. Singh arranged Jagdeep to meet with several staff members of the Punjab Agricultural University. They trained him about agriculture. He spent two months learning about agriculture, fertilizers, seeds and machines. He also learned about subsidiary activities like bee-keeping, mushroom farming and dairy. He received training for 2–3 weeks for each activity.

He completed all the training by taking leave from his job. After this training, Dr. Jagtar suggested him to resign and start full-time farming. His father supported this view.

When Jagdeep resigned, he only had the land. His father was never a farmer. So, Jagdeep requested his in-laws to loan their tractor to him for a season and started cultivating paddy. That crop fetched in INR 136,000 at the end of the first season. He purchased a tractor for INR 120,000. This paltry saving forced him to consider other concurrent activities.

He started bee-keeping. Being cautious, he bought one box for INR 600 and started bee farming. Soon he had 25 boxes, and, by 2002, was producing lots of honey. Sadly, there was no market for honey in his area. He thus decided to leave bee-keeping. However, he made good profit since he sold boxes at 10 times his purchase price.

He thus decided to grow mushrooms. However, there was also no market for mushrooms in Punjab. He had to go to Delhi or Chandigarh to sell his products. This was cumbersome since agriculture was a full time occupation.

Accidentally, he met one of the Route Officers of Nestlé, who encouraged him to consider dairy farming. Jagdeep was not sure, especially after the first two adventures which did not turn out well and left him somewhat disappointed. He decided to start cautiously by buying a buffalo in 2002. Soon, his entrepreneurial skills and profit in selling milk to Nestlé made him a serious visitor to cattle fairs where farmers buy and sell cattle. His constant interactions with Nestlé's veterinary expert's business acumen and sharp eye on the market helped him to increase his herd to 36 buffalos in a span of only two years.

Initially, he encountered some problems. The nearest Nestlé milk collection agency was 8.5 km away. It was not attractive for him to go that far to sell his milk. He sold his milk to the local milkmen at about INR 11 per litre. One day, the Nestlé's local officer persuaded him to come to the agency where he found his milk was of better quality and received INR 14.75 per litre. He said, "This opened my eyes as I quickly calculated if I sell them 10 quintals per day I may get my feed cost back".

Then he met some more officers from Nestlé. They encouraged him to switch to cows. They took him to visit cow farms and tried to convince him of the benefits. He still was not convinced. Nestlé then encouraged him to take 15 days' training at GADVASU. It was there that he could shred his inhibitions about cows. Now he reflects: "I was influenced by what people used to say about cow farming, most of which were based on beliefs and not reasons. After the training, I understood it was all about nutrition and feed, nothing else".

Once he decided to be a dairy farmer, Nestlé helped him to make plans. The company further helped him to get loan of INR 3.6 million from HDFC Bank. From this loan, he made one cowshed and installed two milking machines. He sold all his buffalos and bought 36 cows. Nestlé helped him in the selection and purchase of better yielding cows. This was in 2004.

By the end of 2007, his daily milk production reached at about 900 L. He found it difficult to achieve economy of scale, and was not meeting his own targets. Nestlé assisted him with herd management and other associated issues. He regularly attended dairy-related meetings organized by Nestlé. But still "despite my best intentions and hard work, I was not able to increase my business". On the contrary, because of lack of understanding of cow cycle and lack of modernisation, his milk yield declined to 500 L by mid-2008.

Nestlé noted his concerns and sent him along with other farmers to USA to see and study modern dairy farms. He says:

> We saw, how they manage parlours; feed such huge number of cows and make silage. A lot of technical advice and guidance was given by them with the help of Nestlé. We also learned management of cows and milk production. After coming back from the US, I constructed my own parlour and feed silage to cows. It was a joy that was beyond mentioning!

He introduced synchronization techniques in the new parlours. He decided to increase the number of cows. In 2009, he approached the State Bank of India for a loan of INR 12,000,000 to expand and modernize his farm. The loan was sanctioned in 2010. He then constructed a modern parlour and purchased 40 additional cows and achieved his target of 1,100 L of milk per day.

Jagdeep does not plan to stop here. He would like to construct 2–3 more sheds after constructing his biogas plant and updating milking parlour and old milking plant. As his business grows, efficient and intensive monitoring will be necessary. For example, he says, 'When 100 cows will be milked simultaneously, it is difficult to watch all of them. Therefore, I am left with two alternatives: either conduct milk test and systemic cell count for each cow (this takes 3–5 minutes for each) or to introduce a completely automatic computerized system for cow's yield monitoring.'

Jagdeep says that yield of an animal depends on its systemic cell count. If it goes beyond a certain value both animal's health and milk quality is affected. He thus plans to install fully automatic system at a cost of INR 2,500,000. This system will monitor details of all cows, their body temperature, systemic cell count, heat direction, yields etc. By analyzing these data and planning accordingly, he is confident that milk production will increase further. He told us, "Now I have 168 cows. But, for many of them, I am not sure when they come to 'heat' stage or when they miss it. By this system it will be easy to find out and hence will be easy to manage".

On the strength of this system, he is planning to increase his dairy size to 250 in 2012. He already has prepared the shed to accommodate this number and has constructed two biogas plants. The construction of a new milk parlour is also complete. He appreciates the technical support, inspiration and the culture of quality consciousness of the company. He is continuing to receive advice from the company to identify better cow breeds to improve his herd regularly.

Large scale milk farming is complex. His family consists of his parents, wife and two children. He has 13 permanent employees and equal number of temporary labourers. Primarily due to the training he received from Nestlé, he emphasizes quality management in all aspects of his work. Jagdeep acknowledges role of his wife by saying that, "if labourers gets meal in time, it contributes to the success of the farm". Up to 2011, his wife used to cook for 13–30 persons. However, with changing times, even his father, a highly traditional man, has advised him to hire a cook and receives kitchen gas supply from his modern biogas plants.

Thus, the small seed that was sown by Nestlé has now grown. Jagdeep himself has transformed into a smart businessman. He and his wife now have 67 (female) calves. This is no mean achievement, since a good quality cow costs about INR 85,000 at present. The breed of these in-house calves is now carefully monitored by his wife. The small dairy business that he hesitatingly started in 2002 has now become a major money-spinner.

Jagdeep now says that without the chance meeting with Nestlé in 2002, followed by consistent help and good advice, his story and life would have most likely been very different.

Glossary

Barseem Also called Egyptian or Mediterranean clover, grown as a forage crop and green-manure plant in alkaline soils

Deshi Also spelled desi, native, local, indigenous

Dodi Also known as dudhias, these are milk buyers and traders in charge of collecting milk from farmers and delivering it directly to customers. Such merchants operate in rural, urban and peri-urban areas

Quintal Also called centner, it a historical unit equivalent to 100 kg and it is used a standard measurement of mass for agricultural products

Tona Magical healing rites performed by a shaman or a person familiar with enchantments, sorcery, spells and charms

Octroi Taxes levied on the entry of goods into local areas of consumption, use or sale therein and imposed directly by the municipality and local bodies in India

A. K. Biswas et al., *Creating Shared Value*, SpringerBriefs on Case Studies of Sustainable Development, DOI: 10.1007/978-3-319-01463-0, © The Author(s) 2014

Further Readings

Agarwal B (1983) Mechanization in Indian agriculture: an analytical study based on the Punjab. Delhi School of Economics, New Delhi

Akram-Lodhi AH (2008) (Re) imagining agrarian relations? The world development report 2008. Agriculture for development. Dev Change 39(6):1145–1161

Alderman H (2005) Linkages between poverty reduction strategies and child nutrition: an asian perspective. Econ Polit Weekly 40(46):4837–4842

Bhagwati JN, Desai P (1970) India: planning for industrialisation. Oxford University Press, London

Bhalla GS (1983) Green revolution and the small peasant: a study of income distribution among Punjab cultivators. Concept Pub. Co., New Delhi

Birthal PS, Joshi PK, Gulati A (2005) Vertical coordination in high-value food commodities: implications for smallholders. MTID Discussion paper, No. 85. International Food Policy Research Institute, Washington, DC

Boserup E (1970) Women's role in economic development. St. Martin's Press, New York

Carroll AB (1999) Corporate social responsibility evolution of a definitional construct. Bus Soc 38(3):268–295

Census-2011 (2011). Punjab population census data 2011, India. Available via DIALOG. http://www.census2011.co.in/census/state/punjab.html

Chand R (2004) Dairying in india: experiences and development prospects. National Centre for Agricultural Economics and Policy Research, New Delhi

Chaudhri DP, Dasgupta AK (1985) Agriculture and the development process: a study of punjab. Croom Helm, London

Christiansen N (2008) Creating shared value through basic business strategy, Development Outreach. World Bank, Washington, DC

Candler W, Kumar N (1998) The dairy revolution—the impact of dairy development in india and the world banks' contribution. World Bank, Washington DC

Deep G (2002) The octroi debate: the case of Punjab. In: Singh UB (ed) Revitalised urban administration in India: strategies and experiences. Kalpaz Publications, New Delhi, pp 265–275

De Long B (2003) India since independence: an analytic growth narrative. In D. Rodrik (ed) In search of prosperity: analytic narratives on economic growth. Princeton University Press, Princeton, pp. 184–204

Delgado C, Narrod CA, Tiongco MM (2003) Policy, technical, and environmental determinants and implications of the scaling-up of livestock production in four fast-growing developing countries: a synthesis. Phase I. An IFPRI-FAO Project. Available via DIALOG. http://www.fao.org/WAIRDOCS/LEAD/X6170E/x6170e00.htm#Contents

Dhaliwal KS (2003) Partnership of nestlé with milk producers of india and pakistan. Paper presented at the international workshop on agricultural diversification and vertical integration in south asia. FICCI, ICRISAT and IFPRI, New Delhi, India, 5–6 Nov 2003

Diaz-Alejandro CF (1965) On the import intensity of import substitution. Kyklos 3(3):495–509

A. K. Biswas et al., *Creating Shared Value*, SpringerBriefs on Case Studies of Sustainable Development, DOI: 10.1007/978-3-319-01463-0, © The Author(s) 2014

Dyson T, Moore M (1983) On kinship structure, female autonomy, and demographic behaviour in india. Popul Develop Rev 9(1):35–60
Ederer P Goldberg RA (2007) Nestlé water management strategy: a new competitive advantage. EFAS, Wageningen University, The Netherlands
Food and Agriculture Organization (FAO) (1976) Detailed report of the second world food program evaluation mission on operation floor, Indian dairy corporation. The Economic Scene, No. 7. 10–19
Food and Agriculture Organization (FAO) (2005) Breaking ground: gender and food security. Food and Agriculture Organization, Rome
Frankel FR (1971) India's green revolution. Economic gains and political costs. Princeton University Press, Princeton, NJ
Gereffi G, Wyman D (1990) Paths of industrialization: an overview. In: Gereffi G, Wyman D (eds) Manufacturing miracles: paths of industrialization in latin america and east asia. Princeton University Press, Princeton pp. 3–31
Ghosh RN (1977) Agriculture in economic development—with special reference to punjab. Vikas Publishing House, New Delhi
Gill KPS. (n.d.) Endgame in punjab 1988–1993. Available via DIALOG. http://www.satp.org/satporgtp/publication/faultlines/volume1/Fault1-kpstext.htm
Gill MS, Bhatti JS (n.d.) Dairy scenario of punjab. Available via DIALOG. http://pdfa.org.in/website/english_articles/DAIRY%20INDUSTRIES%20IN%20PUNJAB.pdf
Goldberg RA, Herman K (2007) Nestlé's milk district model: economic development for a value-added food chain and improved nutrition. In: Rangan VK, Quelch JA, Herrero G, Barton B (eds) Business solution for global poor: creating social and economic value. John Wiley, New York, pp. 183–189
Gurchuran D (2000) India unbound: a personal account of a social and economic revolution. Alfred A Knopf, New York
Harish KP, Paramjit SJ, Jagrup SS (1999) Terrorism in punjab: understanding grass-roots reality. Har-Anand Publications, New Delhi
Hemme T, Garcia O Saha A (2003) A review of milk production in india with particular emphasis on small scale producers. PPLPI Working Paper No. 2. Food and Agricul-tural Organization, Italy
International Food Policy Research Institute (IFPRI) (2007) Withering punjab agriculture: can it retain its leadership? Report. IFPRI, New Delhi
International Fund for Agricultural Development (IFAD) (n.d.) Impact of Market-oriented production on household food security. Available via DIALOG. www.ifad.org/hfs/learning/in_3.htm
Interview with Surinder Singh, Veterinary and animal science university (GADVASU), Ludhiana, 14 Nov 2011
Interview with Davinder Singh, 17 Nov 2012
Interview with Manjinder Singh, Village Dharamkot, 17 Nov 2012
Joshi M (1993) Combating terrorism in punjab: indian democracy in crisis. Research Institute for the Study of Conflict and Terrorism, London
Karmakr KG, Banerjee GD (2006) Opportunities and challenges in the indian dairy industry. Technical Digest 2006(9):24–27. Retrieved fromhttp://www.nabard.org/fileupload/DataBank/TechnicalDigest/ContentEnglish/issue9td-6.pdf
Kennedy E, Peters P (1992) Household food security and child nutrition: the interaction of income and gender of household head. World Develop 20(8):1077–1085
Kiram R, Sharma A (2011) Corporate social responsibility: a corporate strategy for new business opportunities. J Int Bus Ethics 4(1):10–17
Kohli A (2004) State-directed development: political power and industrialization in the global periphery. Cambridge University Press, Cambridge
Kulkarni KG, Meister KJ (2008) Trouble with import substitution and protectionism: a Case of indian economy. Int J Econ Res 5(1):35–37

Magaro MM (2010) Two birds, one stone: achieving corporate social responsibility through the stakeholder-primacy norm. Indiana Law J 85(3):1149–1167
McGuire, J., and Popkin, B.M. (1990). Beating the Zero-sum Game: Women and Nutrition in the Third World. Food and Nutrition Bulletin, 11(4), 38–63
McWilliams A, Siegel D (2000) Corporate social responsibility and financial performance: correlation or misspecification? Strategic Manage J 21(5):603–609
Meinzen-Dick R, Behrman J, Menon P, Quisumbing A (2011) Gender: a key dimension linking agricultural programs to improved nutrition and health. Paper presented in 2020 Conference Brief, Feb 9, 2011
Metzger L, Nunnenkamp P, Mahmoud TO (2008) Does corporate aid really help fighting worldwide poverty? a case study of nestlé's aid allocation. Kiel Institute of World Economy. Working paper. Kiel Institute of World Economy, Kiel, Germany
Nestlé Alimentana Company (1975). Nestlé in the developing counties. Nestlé, Vevey
Nestlé H (1857) Nestlé mission and vision. Available via DIALOG. http://www.Nestlé.co.za/aboutus/Pages/MissionVision.aspx
Nestlé (1993) Nestlé in India 1962–1992. Booklet. Nestlé, Vevey
Parvesh S (2011) Punjab females less educated than males: SSA. Times of india (11 Nov) Available via DIALOG. http://articles.timesofindia.indiatimes.com/2011-11-7/india/30369061_1_literacy-rate-males-punjab-education-department
Portney PR (2005) Corporate social responsibility. An economic and public policy perspective. In: Bruce L, Robert N, Richard H (eds) Environmental protection and the social responsibility of firms—perspectives from law, economics, and business. RFF Press Book and Resources for the Future, Washington, pp. 107–131
Porter ME, Kramer MR (2006) The link between competitive advantage and corporate social responsibility. Harvard Bus Rev 84(12):78–92
Rahman O (n.d.) Food security, poverty and women: lessons from rural asia. IFAD, Rome. Available via DIALOG. www.ifad.org/hfs/thematic/rural/rural_toc.htm
Ravi K, Sharma A (2011) A corporate strategy for new business opportunities. J Int Bus Ethics 4(1):10–17
Roberts SL (2001) Women: the key to food security. Paper presented at the International Congress on Dietetics, Chicago, IL
Rodrik D (2005) Growth strategies. Handbook of Economic Growth 1:967–1014
Sandhu TS (n.d.) Parasitism: a cancer on production. Food Specialities Limited, Moga
Sandhu TS (n.d.) Development of milk shed area. Food Specialities Limited, Moga
Sandhu TS (n.d.) Education and information on animal husbandry techniques for rural women, why and how. Food Specialities Limited, Moga
Sandhu TS (1978) Agro-climatic and other situations. Yearbook. Food Specialities Limited, Moga
Sandhu TS (1980a) Impact of moga milk plant on agricultural and rural development. Food Specialities Limited, Moga
Sandhu TS (1980b) Anatomy of milk sales. Food Specialities Limited, Moga
Sandhu TS (1980c) Human factor and milk forecasting. Food Specialities Limited, Moga
Sandhu TS, Dhaliwal KS (1980) Rural poverty. Food Specialities Limited, Moga
Sandhu TS (1981a) Extent of full employment generated by moga milk plants in its milk shed area. Moga: Food Specialities Limited
Sandhu TS (1981b) Evaluation of development activities of moga milk plant. Food Specialities Limited, Moga
Sandhu TS (1981c) Inputs for milk procurement. Food Specialities Limited, Moga
Sandhu TS (1981d) Prosperity? Food Specialities Limited, Moga
Sandhu TS (1982a) Alcoholism. Rural development study. Food Specialities Limited, Moga
Sandhu TS (1982b) Fulfillment of the concepts of income, output and employment by moga milk plant. Food Specialities Limited, Moga
Sandhu TS (1982c) Efficient methods of clean milk production. Food Specialities Limited, Moga

Sandhu TS (1984) Constraints in milk production and their removal. Food Specialities Limited, Moga

Sandhu TS (1985a) Challenges for milk procurement and extension workers. Food Specialities Limited, Moga

Sandhu TS (1985b) Milk development of milk shed area. Food Specialities Limited, Moga

Sandhu TS (1986) Feeding farm ruminants. Food Specialities Limited, Moga

Sandhu TS (1987) Perspective planning for milk pricing in the private sector dairies. Food Specialities Limited, Moga

Sandhu TS (1988a) Golden and glorious period. (Unprinted memoirs). Food Specialities Limited, Moga

Sandhu TS. (1988b) Genesis, goals and roles and benefits of field days organised in the milk district of food. Food Specialities Limited, Moga

Sandhu TS (1988c) The story of FSL. Food Specialities Limited, Moga

Satyapal D, Chopra VD, Bakaya RM (2000) Terrorism in punjab. Gyan Books, New Delhi

Shukla RK, Brahmankar SD (1999) Impact evaluation of operation flood on the rural dairy sector. National Council of Applied Economic Research, New Delhi

Sharma VP, Vir Singh R, Staal S, Delgado C (2002) Critical issues for poor in the India dairy sector on the threshold of a new era. Livestock Industrialisation Project (Phase). An IFPRI-FAO Project, Food and Agriculture Organization, Rome

Singa OP (2007) Agro-industries characterization and appraisal: dairy in india. Dairy Farmers' Organization, Anand, India. Working Paper No. 21. Food and Agricultural Organization, Rome

Singh A, Jassi J (1992) The bleeding punjab: as report to the nation. Joint publication of surkh rekha and inqulib jantak. Available via DIALOG. http://iref.homestead.com/khalistan.html

Singh P (2008) Causes and consequences of terrorism in punjab: a rationalist perspective. Institute for Defence Studies and Analyses, New Delhi. Available via DIALOG. http://www.idsa.in/idsastrategiccommentsCausesandconsequencesofTerrorisminPunjab_PSingh_180708

Technology Export Development Organization (TEDO) (n.d.) Milk and diary sector: an overview. Undated. Department of Scientific and Industrial Research. Technology Promotion, Development and Utilization Program. International Technology Transfer Program

Tiganas A, Bacali R, Vazquez-Burguete JL (n.d.) Nestlé and moga district of india: a half century of "Marriage" base on CSR values. Paper. 3rd International Congress on Teaching Cases Related to Public and Nonprofit Marketing

UN Department of Humanitarian Affairs (1988) India-Floods. UNDRO Information Report No. 1. Available via DIALOG. http://reliefweb.int/node/35060

World Bank (1999) India livestock sector review: enhancing growth and development. Allied Publishers, Mumbai

Index

A. K. Biswas et al., *Creating Shared Value*, SpringerBriefs on Case Studies of Sustainable Development, DOI: 10.1007/978-3-319-01463-0, © The Author(s) 2014

MIX
Papier aus verantwortungsvollen Quellen
Paper from responsible sources
FSC® C105338

If you have any concerns about our products,
you can contact us on
ProductSafety@springernature.com

In case Publisher is established outside the EU,
the EU authorized representative is:
Springer Nature Customer Service Center GmbH
Europaplatz 3, 69115 Heidelberg, Germany

Printed by Libri Plureos GmbH
in Hamburg, Germany